Subspace Learning of Neural Networks

AUTOMATION AND CONTROL ENGINEERING

A Series of Reference Books and Textbooks

Series Editors

FRANK L. LEWIS, Ph.D.,
Fellow IEEE, Fellow IFAC
Professor
Automation and Robotics Research Institute
The University of Texas at Arlington

SHUZHI SAM GE, Ph.D.,
Fellow IEEE
Professor
Interactive Digital Media Institute
The National University of Singapore

Subspace Learning of Neural Networks, *Jian Cheng Lv, Zhang Yi, and Jiliu Zhou*

Reliable Control and Filtering of Linear Systems with Adaptive Mechanisms, *Guang-Hong Yang and Dan Ye*

Reinforcement Learning and Dynamic Programming Using Function Approximators, *Lucian Buşoniu, Robert Babuška, Bart De Schutter, and Damien Ernst*

Modeling and Control of Vibration in Mechanical Systems, *Chunling Du and Lihua Xie*

Analysis and Synthesis of Fuzzy Control Systems: A Model-Based Approach, *Gang Feng*

Lyapunov-Based Control of Robotic Systems, *Aman Behal, Warren Dixon, Darren M. Dawson, and Bin Xian*

System Modeling and Control with Resource-Oriented Petri Nets, *Naiqi Wu and MengChu Zhou*

Sliding Mode Control in Electro-Mechanical Systems, Second Edition, *Vadim Utkin, Jürgen Guldner, and Jingxin Shi*

Optimal Control: Weakly Coupled Systems and Applications, *Zoran Gajić, Myo-Taeg Lim, Dobrila Skatarić, Wu-Chung Su, and Vojislav Kecman*

Intelligent Systems: Modeling, Optimization, and Control, *Yung C. Shin and Chengying Xu*

Optimal and Robust Estimation: With an Introduction to Stochastic Control Theory, Second Edition, *Frank L. Lewis, Lihua Xie, and Dan Popa*

Feedback Control of Dynamic Bipedal Robot Locomotion, *Eric R. Westervelt, Jessy W. Grizzle, Christine Chevallereau, Jun Ho Choi, and Benjamin Morris*

Intelligent Freight Transportation, *edited by Petros A. Ioannou*

Modeling and Control of Complex Systems, *edited by Petros A. Ioannou and Andreas Pitsillides*

Wireless Ad Hoc and Sensor Networks: Protocols, Performance, and Control, *Jagannathan Sarangapani*

Stochastic Hybrid Systems, *edited by Christos G. Cassandras and John Lygeros*

Hard Disk Drive: Mechatronics and Control, *Abdullah Al Mamun, Guo Xiao Guo, and Chao Bi*

Autonomous Mobile Robots: Sensing, Control, Decision Making and Applications, *edited by Shuzhi Sam Ge and Frank L. Lewis*

Automation and Control Engineering Series

Subspace Learning of Neural Networks

Jian Cheng Lv

Sichuan University
Chengdu, People's Republic of China

Zhang Yi

Sichuan University
Chengdu, People's Republic of China

Jiliu Zhou

Sichuan University
Chengdu, People's Republic of China

CRC Press
Taylor & Francis Group
Boca Raton   London   New York

CRC Press is an imprint of the
Taylor & Francis Group, an **informa** business

CRC Press
Taylor & Francis Group
6000 Broken Sound Parkway NW, Suite 300
Boca Raton, FL 33487-2742

© 2011 by Taylor and Francis Group, LLC
CRC Press is an imprint of Taylor & Francis Group, an Informa business

No claim to original U.S. Government works

Printed in the United States of America on acid-free paper
10 9 8 7 6 5 4 3 2 1

International Standard Book Number: 978-1-4398-1535-9 (Hardback)

Visit the Taylor & Francis Web site at
http://www.taylorandfrancis.com

and the CRC Press Web site at
http://www.crcpress.com

Dedication

To all of our loved ones

Preface

Principal component analysis (PCA) neural networks, minor component analysis (MCA) neural networks and independent component analysis (ICA) neural networks can approximate a subspace of input data by learning. These networks inspired by biology and psychology provide a novel way for parallel online computation of a subspace. An input of these neural networks can be used at once so that they can enable fast adaptation in a nonstationary environment. Although these networks are almost linear neural models, they have found many applications, including applications relating to signal and image processing, video analysis, data mining, and pattern recognition.

The learning algorithms of these neural networks play a vital role in subspace learning. These subspace learning algorithms make these networks learn low-dimensional linear and multilinear models in a high-dimensional space, wherein specific statistical properties can be well preserved. The book will be mainly focused on the convergence analysis of these subspace learning algorithms and the ways to extend the use of these networks to fields such as biomedical signal processing, biomedical image processing, and surface fitting to name just a few.

A crucial issue of concern in a practical application is the convergence of the subspace learning algorithms of these neural networks. The convergence of these algorithms determines whether these applications can be successful. The book will analyze the convergence of these learning algorithms by mainly using discrete deterministic time (DDT) method. To guarantee their nondivergence, invariant sets of some algorithms will be obtained and global boundedness of some algorithms is studied. Then, the convergence conditions of these algorithms will be derived. Cauchy convergence principle and inequalities analysis method, and so on, will be used rigorously to prove the convergence. Furthermore, the book establishes a relationship between an SDT algorithm and the corresponding DDT algorithm by using block algorithms. This not only can overcome the shortcomings of DDT method, but also can get a good convergence and accuracy in practice. Finally, the chaotic and robust properties of some algorithms will also be studied. These results obtained lay the sound theoretical foundation of these networks and guarantee the successful applications of these algorithms in practice.

The book not only benefits the researcher of subspace learning algorithms, but also improves the quality of data mining, image processing, and signal processing. Besides its research contributions and applications, the book could also serve as a good example for pushing the latest technologies in neural networks to some application community.

Scope and Contents of This Book

This book provides an analysis framework for convergence analysis of subspace learning algorithms of neural networks. The emphasis is on the analysis method, which can be generalized to the study of other learning algorithms. Our work builds a theoretical understanding of the convergence behavior of some subspace learning algorithms through the analysis framework. In addition, this book uses real-life examples to illustrate the performance of learning algorithms and instructs readers on how to apply them to practical applications. The book is organized as follows.

Chapter 1 provides a brief introduction to linear neural networks and subspace learning algorithms of neural networks. Some frequently used notations and preliminaries are given. Basic discussions on the methods for convergence analysis are presented which should lay the foundation for subsequent chapters.

In the following chapters, convergence of subspace learning algorithms is analyzed to lay the theoretical foundation for successful applications of these networks. In Chapter 2, the convergence of Oja's and Xu's algorithms with constant learning rates is studied in detail. The global convergence of Oja's algorithm with the adaptive learning rate is analyzed in Chapter 3. In Chapter 4, the convergence of Generalized Hebbian Algorithm (GHA) with adaptive learning rates is studied. MCA learning algorithms and the Hyvärinen-Oja's ICA learning algorithm are analyzed in Chapters 5 and 6, respectively. In Chapter 7, chaotic behaviors of subspace learning algorithms are presented.

Some problems concerning a practical application are discussed in chapters 8, 9, 10, 11, and some real-life examples are given to illustrate the performance of these subspace learning algorithms.

The contents of this book are mainly based on our research publications on this subject, which over the years have accumulated into a complete and unified coverage of the topic. It will serve as an interesting reference for postgraduates, researchers, and engineers who may be keen to use these neural networks in applications. Undoubtedly, there are other excellent works in this area, which we hope to have included in the references for the readers. We should also like to point out that at the time of this writing, many problems relating to subspace learning remained unresolved, and the book may contain personal views and conjecture of the authors that may not appeal to all sectors of readers. To this end, readers are encouraged to send us criticisms and suggestions, and we look forward to discussion and collaboration on the topic.

Acknowledgments

This book was supported in part by the National Science Foundation of China under grants 60971109 and 60970013.

Jian Cheng Lv
Zhang Yi
Jiliu Zhou

January 2010

List of Figures

List of Tables

Contents

1

Introduction

1.1 Introduction

Subspace leaning neural networks provide a parallel online automated learning of low-dimensional models in a nonstationary environment. It is commonly known that automated learning of low-dimensional linear or multilinear models from training data has become a standard paradigm in computer vision [54, 55]. Thus, these neural networks used to learning a low-dimensional model have many important applications in computer vision, such as structure from motion, motion estimation, layer extraction, objection recognition, and object tracking [18, 19, 50, 56, 55].

Subspace learning algorithms used to update the weights of these networks pay a vital important role in applications. This book will focus on the subspace learning algorithms of principal component analysis (PCA) neural networks, minor component analysis (MCA) neural networks, and independent component analysis (ICA) neural networks. Generally, they are linear neural networks.

1.1.1 Linear Neural Networks

Inspired by biological neural networks, a simple neural model is designed to mimic its biological counterpart, the neuron. The model accepted the weighted set of input x responds with an output y, as shown in Figure 1.1. The vital effect of synapse between neurons is presented by the weights w. The mathematical model of a linear single neuron is as follows:

$$y(k) = w^T(k)x(k), (k = 0, 1, 2, \ldots), \qquad (1.1)$$

where $y(k)$ is the network output, the input sequence

$$\{x(k) \,|\, x(k) \in R^n (k = 0, 1, 2, \ldots)\} \qquad (1.2)$$

is a zero-mean stochastic process, each $w(k) \in R^n (k = 0, 1, 2, \ldots)$ is a weight vector.

Consider input output relation

$$y(k) = W^T x(k), \qquad (1.3)$$

1

where

$$y(k) = [y_1(k), y_2(k), \ldots, y_m(k)]^T$$

and

$$W = \begin{bmatrix} w_{11} & w_{12} & \cdots & w_{1m} \\ w_{21} & w_{22} & \cdots & w_{2m} \\ \cdots & \cdots & \cdots & \cdots \\ w_{n1} & w_{n2} & \cdots & w_{nm} \end{bmatrix}.$$

This is a multineuron linear model, as shown in Figure 1.2.

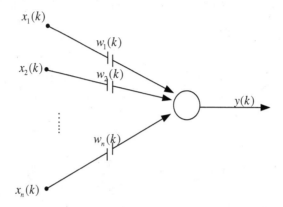

FIGURE 1.1
A single neuron model.

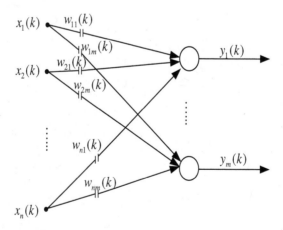

FIGURE 1.2
A multineuron model.

These linear networks have been widely studied [25, 28, 70] and used for many fields involving signal processing [66], prediction [80], associative memory [167], function approximation [180], power system [15, 153], chemistry [196], and so on. The weight of these neural networks, which represents the strength of the connection between neurons, pays a very important role in the different applications. A variety of learning algorithms are used to update the weight so that the networks have the different applications with the corresponding weight. For instance, the adaptive linear neuron (ADALINE) network is one of the most widely used neural networks in practical applications, which was introduced by Widrow and M. Hoff in 1960 [181]. The least mean square (LMS) algorithm is used to update the weight so that ADALINE can be used as a adaptive filter [75].

In this book, these linear networks are used to extract the principal components, or minor components, or independent components from input data. It is required that these networks must approximate a low-dimensional model.

1.1.2 Subspace Learning Algorithms

Subspace learning algorithms are used to update the weights of these linear networks so that these networks intend to approximate a low-dimensional linear model. Generally, an original stochastic discrete time (SDT) algorithm is formulated as

$$w(k+1) = w(k) \pm \eta(k) \triangle w(k), \qquad (1.4)$$

where $\eta(k)$ is learning rate and $\triangle w(k)$ determines the change at time k.

This book will mainly discuss the following subspace learning algorithms: PCA learning algorithms, MCA learning algorithms, and ICA learning algorithms.

1.1.2.1 PCA Learning Algorithms

Principal component analysis (PCA) is a traditional statistical technique in multivariate analysis, stemming from the early work of Pearson [141]. It is closely related to Karhunen-Loève (KL) transform, or the Hotelling transform [82]. The purpose of PCA is to reduce the dimensionality of a given data set, while retaining as much as possible of the information present in the data set.

Definition 1.1 *A vector is called the first principal component direction if the vector is along the eigenvector associated with the largest eigenvalue of the covariance matrix of a given data set, and a vector is called the second principal component direction if the vector is along the eigenvector associated with the second largest eigenvalue of the covariance matrix of a given data set, and so on.*

Definition 1.2 *Principal components are the variances that are obtained by projecting the given data onto the principal component directions.*

Definition 1.3 *a subspace is called a principal subspace if the subspace is spanned by the first few principal component directions.*

Generally, PCA has two steps. First, principal component directions must be found. In these directions, the input data have the largest variances, as shown in Figure 1.3. Then, principal components are obtained by projecting the input data into the principal directions. The principal components are uncorrelated and ordered so that the first few retain most of the variation present in all of the original data set.

PCA has a wide variety of different applications, as well as a number of different derivations [17, 24, 29, 93, 150], such as Kernel PCA [42, 99], Probabilistic PCA [170, 171, 207], Local PCA [116, 179], Generalized PCA [173], L1-PCA [104], Blockwise PCA [41, 133], Sparse PCA [208], Topological PCA [148], and Bayesian PCA [176, 112].

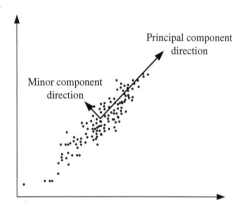

FIGURE 1.3
Principal component direction and minor component direction in a two-dimensional data set.

Traditionally, some classical methods such as QR decomposition or SVD algorithm are used to find the principal component directions. These methods are not practical to handle large data sets because of huge covariance matrix [73, 179, 189]. PCA learning algorithms offer an advantage that they can adapt online to the data and no explicit computation of the covariance matrix.

PCA learning algorithms are used to update the weights of linear neural networks so that the weights converge to the principal component directions that span a principal subspace. Then, these networks can be used to extract the principal components without calculating the covariance matrix in advanced. PCA neural networks have many applications, including applications relating to feature extraction [152, 158], fault detection [90, 109], face classification [114, 166], gender recognition [53], image compression [51], text classification [113], forecasting [118], and protein structure analysis [64].

In [134], Oja used a simple linear single neural network (Figure 1.1) to extract the principal components from the input data. Based on the well-known Hebbian learning rule, Oja proposed a PCA learning algorithm to make the weights of the network converge to the first principal component direction. Thus, this network, under Oja's algorithm, is able to extract principal components from input data adaptively. The results have been useful in online data processing applications.

Stemming from Oja's algorithm [134], many PCA algorithms have been proposed to update the weights of these networks [1, 7, 44, 103, 131, 139, 156, 184]. Among the algorithms for PCA, Oja's algorithm and Xu's LMSER algorithm are commonly used in these applications. Several other algorithms for PCA are related to these basic procedures [32]. In [156], Sanger proposed the GHA learning algorithm, which allows a multineuron linear neural network (Figure 1.2) to extract a selected number of principal components. In [103], Kung et al. proposed the application of the APEX learning algorithm to a neural network with lateral connections to extract all the desired principal components in parallel. In [44], Cichocki et al. proposed the CRLS algorithm, which combines the advantages of both the SAMH algorithm [1] and the RLS-PCA algorithm [7] to extract principal components sequentially. There are many other extensions, see, for examples, [31, 37, 81, 91, 135, 175, 179, 203, 204].

A review of PCA learning algorithms can be found in [30].

1.1.2.2 MCA Learning Algorithms

Minor component was explicitly introduced by Xu in [185], which is the counterpart of principal component. Similar to PCA, some definitions are as follows.

Definition 1.4 *A vector is called the first minor component direction if the vector is along the eigenvector associated with the smallest eigenvalue of the covariance matrix of a given data set, and a vector is called the second minor component direction if the vector is along the eigenvector associated with the second smallest eigenvalue of the covariance matrix of a given data set, and so on.*

Definition 1.5 *A subspace is called a minor subspace if the subspace is spanned by the first few minor component directions.*

Figure 1.3 shows a minor component direction in a 2D data set. Minor components obtained by projecting the input data onto the minor component directions are often regarded as unimportant or noise. However, it is found that MCA has many applications in diverse areas, such as total least squares (TLS) [71, 72], moving target indication [100], clutter cancellation [9], computer vision [47], curve and surface fitting [137, 185], digital beamforming [74], frequency estimation [128], and bearing estimation [157].

MCA neural networks can extract the minor components from input data set without calculating the covariance matrix in advance. The corresponding learning algorithms are proposed to update the net weights, such as the Xu's algorithm[185], the OJA+'s algorithm [135], the Luo-Unbehauen-Cichocki's algorithm [119], the Feng-Bao-Jiao's learning algortihm [67], the EXIN's algorithm [48], Feng's OJAm algorithm [68], and Fast MCA algorithm [12].

An overview of MCA learning algorithms can be found in [130].

1.1.2.3 ICA Learning Algorithms

Independent component analysis (ICA) [49] can estimate the independent components from the mixed input signal. Its goal is to find a subspace in which a given input signals can be decomposed into groups of mutually independent random signals. The essential difference between ICA and PCA is that PCA decomposes a set of signals into a set of uncorrelated, ordered components, whereas ICA decomposes a set of signal mixtures into a set of independent, unordered signals [164]. This implies that PCA transforms the input signal $x = (x_1, x_2)$ with joint probability density function (PDF) $p_x(x_1, x_2)$ into a new set of uncorrelated components $y = (y_1, y_2)$ so that

$$E[y_1 y_2] = E[y_1] E[y_2].$$

In contrast, ICA transforms the signal mixtures $x = (x_1, x_2)$ with PDF $p_x(x_1, x_2)$ into a new set of independent signals $y = (y_1, y_2)$ so that

$$p_{y_1, y_2}(y_1, y_2) = p_{y_1}(y_1) p_{y_2}(y_2).$$

Actually, PCA does more than simply find a transformation of the input signals such that the extracted principal components are uncorrelated. PCA orders the extracted principal components according to their variances, so that components associated with high variance are deemed more important than those with low variance. In contrast, ICA is essentially blind to the variance associated with each extracted signal. The only assumption of ICA method is the source signals are statistically independent and nonGaussian. ICA can extract components that are both statistically independent and nonGaussian even if very little is known about the nature of those source signals.

The data analyzed by ICA could originate from many application fields as diverse as speech processing, brain imaging, electrical brain signals, telecommunications, and stock market prediction. In many cases, these data are given as a set of signal mixtures, the term blind source separation is used to characterize this problem. The main applications of ICA are in blind source separation [21, 57, 94, 102, 108], feature extraction [13, 96, 155], and blind deconvolution [59, 159, 168].

ICA neural networks enable fast adaptation in a stochastic environment since the inputs can be used in their algorithms at once [84]. ICA learning algorithms are used to update the weights of ICA neural networks so that the weights converge to the subspace we expected. These networks have been extensively studied and used in many fields; see examples [3, 20, 22, 23, 97, 106, 108, 138, 169].

1.1.3 Outline of This Book

The rest of this chapter is organized as follows. In Section 2, the methods for convergence analysis is introduced and the relationship between SDT algorithm and the corresponding DDT algorithm is given in Section 3. Section 4 presents some notations and preliminaries.

In Chapter 2, the convergence of Oja's and Xu's algorithms with constant learning rates is studied in detail. Some invariant sets are obtained and local convergence is proven rigorously. The most important contribution is the convergence analysis framework of deterministic discrete time (DDT) method is established in this chapter.

In Chapter 3, an adaptive learning rate is provided for the subspace learning algorithms. The global boundedness of Oja's algorithm with the adaptive learning rate is analyzed first. Then, the global convergence conditions are obtained.

In Chapter 4, the convergence of Generalized Hebbian Algorithm (GHA) is studied via the DDT method. GHA can extract multiple principal component directions in parallel. In this chapter, mathematical induction is used to prove the global convergence of GHA with adaptive learning rates.

In Chapter 5, a stable MCA algorithm is proposed, and its dynamics of the proposed algorithm is interpreted indirectly by using DDT method.

In Chapter 6, the Hyvärinen-Oja's ICA learning algorithm is discussed. The corresponding DDT algorithms are extended to the block versions of the original SDT algorithms. The block algorithms not only establish a relationship between the SDT algorithms and the corresponding DDT algorithms, but also can get a good convergence speed and accuracy in practice.

In Chapter 7, chaotic behaviors of subspace learning algorithms are presented. It will be shown that some algorithms have both stability and chaotic behavior under some conditions. This chapter explores the chaotic behavior of a class of ICA algorithms. The conditions for stability and chaos of the algorithm are derived.

In Chapter 8, a biologically plausible PCA model is proposed to adaptively determine the number of principal component directions. The number adaptively approximates the intrinsic dimensionality of the given data set by using an improved GHA. Some simulations are conducted to verify the theoretical results.

In Chapter 9, an adaptive learning rate is proposed to guarantee the global convergence of Oja+'s MCA algorithm. Then, a multi-block-based MCA method is used for nonlinear surface fitting. The method shows good performance in terms of accuracy, which is confirmed by the simulation results.

In Chapter 10, an ICA algorithm is proposed for extracting fetal electrocardiogram (FECG). Theorectical analysis of the algorithm is also given in this chapter.

Finally, in Chapter 11, a rigid medical image registration method is introduced. The proposed method accomplishes registration by simply aligning the first principal component direction and centroids of feature images. Since a PCA neural network is used to compute the first principal component direction of feature images, the registration is simple and efficient.

1.2   Methods for Convergence Analysis

The learning algorithms of neural networks play an important role in their practical applications. The convergence of learning algorithms determines whether these applications can be successful. However, it is very difficult to directly study the convergence of these learning algorithms, because they are described by stochastic discrete time (SDT) algorithms [202]. Generally, two methods are used to analyze the convergence of learning algorithms. Traditionally, based on some stochastic approximation theorems [14, 35, 117], an original SDT algorithm is transformed into a corresponding deterministic continuous time (DCT) algorithm. The convergence of the SDT algorithm can be indirectly interpreted by studying the corresponding DCT algorithm. This approach is called the DCT method. Another method is called the deterministic discrete time (DDT) method. The DDT method transforms a SDT algorithm into a corresponding DDT algorithm by taking the conditional expectation on both sides of the SDT algorithm. The DDT algorithm characterizes the average evolution of the original SDT algorithm. The DDT algorithm preserves the discrete time nature of the original algorithm. The convergence of the DDT algorithm can shed some light on the dynamical behaviors of the original SDT algorithm.

1.2.1   DCT Method

Traditionally, based on some stochastic approximation theorems [14, 117], an original SDT algorithm is transformed into a corresponding deterministic continuous time (DCT) algorithms; that is, the algorithm (1.4) is transformed into the following differential equation (1.5):

$$\dot{w} = \triangle w(k). \tag{1.5}$$

The convergence of the algorithm (1.4) can be indirectly interpreted by studying the corresponding DCT algorithm (1.5). Because of mathematical achievements of the differential equation, the method has been widely used; see for examples, [4, 36, 135, 136, 154, 162, 172, 192, 199].

In order to transform an SDT algorithm to a DCT algorithm, some restrictive conditions must be applied. One of these conditions is that

$$\eta(k) > 0, \quad \sum_k \eta(k) = \infty, \quad \sum_k (\eta(k))^2 < \infty,$$

where $\eta(k)$ is the learning rate of the SDT learning algorithm (1.4). This condition implies that the learning rate $\eta(k)$ must converge to zero. However, it is unrealistic in practical applications, because of computational round-off limitations and tracking requirements, whenever the step increases considerably, small but nonzero constant values of learning rate have to be employed [210]. If the learning rate approaches zero, clearly, it could slow down the convergent speed of the algorithm considerably. In addition, the learning rate cannot approach zero too fast; otherwise, the algorithm will stop before it converges to the right direction. More discussions can be found in [39, 195, 202, 210]. Thus, the condition above is unrealistic in practical applications.

1.2.2 DDT Method

To overcome the shortcomings of the DCT method, the deterministic discrete time (DDT) method has been proposed recently to indirectly study the convergence of SDT learning algorithms; see [39, 124, 195, 210]. The DDT method transforms the SDT algorithm (1.4) into the corresponding DDT algorithm as follows:

$$E\{w(k+1)\} = E\{w(k)\} \pm E\{\eta(k) \triangle w(k)\}. \tag{1.6}$$

Then, to guarantee the nondivergence of DDT algorithms, the boundedness of the algorithms must be analyzed firstly. For global convergent algorithms, global boundedness is studied. For local convergent algorithms, invariant sets are obtained to guarantee the local nondivergence. The invariant set is defined as follows.

Definition 1.6 *A compact set $S \subset R^n$ is called an invariant set of the algorithm $w(k + 1) = f(w(k), x(k))$, if for any $w(0) \in S$, the trajectory of the algorithm starting from $w(0)$ will remain in S for all $k \leq 0$.*

An invariant set is interesting because it provides a method to guarantee nondivergence of an algorithm. Clearly, a subset of an invariant set may be not an invariant set. If a set has an invariant subset, it does not imply this set is also an invariant set [195].

Under the conditions of nondivergence, Cauchy convergence principle, inequalities analysis method, and so on, are used to derive the convergence conditions and rigorously prove the convergence of algorithms. The results can shed some light on the original algorithms.

The DDT method does not require the learning rates to converge to zero so that constant learning rates can be used. The DDT algorithms preserve the discrete time nature of the original algorithms and allow a more realistic dynamic behavior of the learning rates [210]. Compared to the DCT method, the DDT method is a more reasonable approach for indirectly studying the convergence of SDT algorithms.

The DDT algorithm characterizes the average evolution of the original SDT algorithm. In a stochastic environment, the stochastic input has a great influence on the result of DDT method, even derailing the convergent conditions of the DDT algorithm [120]. In Chapter 6, we propose a method to avoid the issue. First, nondivergence of an original algorithm is guaranteed in stochastic environment by directly studying the original algorithm. Under this condition, convergence of the algorithm is analyzed via DDT method.

1.3   Relationship between SDT Algorithm and DDT Algorithm

The SDT algorithm (1.4) and the corresponding DDT algorithm (1.6) all preserve the discrete time nature. The DDT algorithm (1.6) can also be presented as

$$\frac{1}{L}\sum_{i=1}^{L}\{w(k+1)\} = \frac{1}{L}\sum_{i=1}^{L}\{w(k)\} \pm \frac{1}{L}\sum_{i=1}^{L}\{\eta(k) \triangle w(k)\}, L \to \infty.$$

Clearly, if $L = 1$, this is the SDT algorithm (1.4). Thus, let L be some constants. We have got some algorithms called block algorithm. The block algorithms not only establish a relationship between an original SDT algorithm and the corresponding DDT algorithm, as shown in Figure 1.4, but also get a compromise between the SDT algorithm and the DDT algorithm.

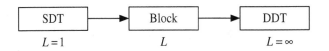

FIGURE 1.4
Relationship between SDT algorithm and DDT algorithm.

The block size L plays an important role in the block algorithms, which can be determined according to the practical applications [129]. For an SDT algorithm and the corresponding block algorithms and DDT algorithm, it is easy to get the following result.

Corollary 1.1 *The invariant sets of an original SDT algorithm are also the invariant sets of the corresponding block algorithms and DDT algorithm, but not vice versa.*

It is easy to see why the stochastic input has a great influence on the result of DDT method. Thus, obtaining the invariant set of the original SDT algorithm is a good solution to the problem. Chapter 6 explains this problem in detail.

1.4 Some Notations and Preliminaries

1.4.1 Covariance Matrix

For the input sequence (1.2), let $C = E\left[x(k)x^T(k)\right]$. It is known that the covariance matrix C is a symmetric nonnegative definite matrix. There exists an orthonormal basis of R^n composed by eigenvectors of C. Let $\lambda_1, \ldots, \lambda_n$ be all the eigenvalues of C, ordered by $\lambda_1 \geq \ldots \geq \lambda_n \geq 0$. Suppose that $\{v_i | i = 1, \ldots, n\}$ is an orthonormal basis in R^n such that each v_i is a unit eigenvector of C associated with the eigenvalue λ_i.

For each $k \geq 0$, $w(k) \in R^n$ can be represented as

$$w(k) = \sum_{i=1}^{n} z_i(k)v_i, \tag{1.7}$$

where $z_i(k)(i = 1, \ldots, n)$ are some constants, and then

$$Cw(k) = \sum_{i=1}^{n} \lambda_i z_i(k)v_i = \sum_{i=1}^{p} \lambda_i z_i(k)v_i. \tag{1.8}$$

1.4.2 Simulation Data

The well-known gray picture of Lenna in Figure 1.5 (left) will be often used for simulations in this book. The 512×512 pixel image is decomposed into 4096 blocks that do not intersect one another, as shown in Figure 1.5 (right). Each block comprises 8×8 pixels, which constructs a 64-dimensional vector. Thus, a vector set with 4096 vectors is obtained. Upon removal of the mean and normalization, randomly selected vectors from this vector set form an input sequence

$$\{x(k)|x(k) \in R^{64}, (k = 1, 2, 3, \ldots)\}. \tag{1.9}$$

To cater to the tracking capability of the adaptive algorithm, we will often use C_k to replace $x(k)x^T(k)$ in (1.4); the C_k is defined as follows [32, 31]:

$$C_k = \beta C_{k-1} + \frac{1}{k}\left(x(k)x^T(k) - \beta C_{k-1}\right), \quad (k > 0) \tag{1.10}$$

where β is a forgetting factor, which ensures that the past data samples are down-weighted. Since each C_k contains some history information of the input

FIGURE 1.5
The original picture of Lenna (left), and the picture divided into 4096 blocks (right).

data, it can accelerate the convergent speed and reduce the oscillation of the algorithm. Generally, we set $\beta = 1$ if $\{x(k)\}$ comes from a stationary process.

In simulations, signal-to-noise ratio (SNR) is calculated to measure the quality of image compression as

$$SNR = 10log_{10} \frac{\sum_{i=1}^{row} \sum_{j=1}^{col} I_{ij}^2}{\sum_{i=1}^{row} \sum_{j=1}^{col} (I_{ij}^2 - \hat{I}_{ij}^2)}, \tag{1.11}$$

where $I_{ij}^2, \hat{I}_{ij}^2$ represent the pixel of original image or reconstructed image, respectively.

1.5 Conclusion

This chapter first provides a brief introduction to linear neural networks and subspace learning algorithms of neural networks. Then, the DCT method and DDT method for convergence analysis are introduced. It can be seen that DDT method is a more reasonable approach for indirectly studying the original SDT algorithm. At the same time, the relationship between a SDT algorithm and the corresponding DDT algorithm is given. Finally, some notations and preliminaries are presented. This chapter lays the foundation for subsequent chapters.

2

PCA Learning Algorithms with Constants Learning Rates

2.1 Introduction

Principal component analysis (PCA) neural networks are useful tools in feature extraction, data compression, pattern recognition, and time series prediction, especially in online data precessing applications. The learning algorithms of PCA neural networks play an important role in their practical applications. Stemming from Oja's algorithm [134], many PCA algorithms have been proposed to update the weights of these networks. Among the algorithms for PCA, Oja's algorithm and Xu's LMSER algorithm are commonly used in practical applications. Several other algorithms for PCA are related to the two basic procedures [32]. In this chapter, we will study the Oja's and Xu's algorithms and establish the theoretical foundation of their applications.

The convergence of the learning algorithms determines whether these applications can be successful. This chapter propose a convergence analysis framework, called DDT method, to study the convergence of Oja's and Xu's algorithms with constant learning rates. The original SDT algorithms are first transformed into the corresponding DDT algorithms. Then, local nondivergence conditions are guaranteed by using the invariant sets. Finally, in the invariant sets, the local convergence properties of the DDT algorithms are discussed via Cauchy principle, inequalities analysis method, and so on.

This chapter is organized in three parts. The first part[1] analyzes the convergence of Oja's algorithm with a constant learning rate via DDT method. In Section 2.2.1, we will briefly discuss the Oja's PCA learning algorithm and the traditional DCT method. Some preliminaries and the DDT formulation will be presented in Section 2.2.2. In Section 2.2.3, invariant sets and convergence will be studied. Simulation results and discussions will be provided in Section 2.2.4. Finally, conclusions will be drawn in Section 2.2.5.

The second part of the chapter discusses the convergence of Xu's LMSER algorithms with a constant learning rate. Section 2.3.1 describes the formulation and preliminaries. In Sections 2.3.2 and 2.3.3, the convergence results are

---

[1]Based on "Convergence Analysis of a deterministic discrete time system of Oja's PCA learning algorithm," by (Zhang Yi, M. Ye, Jian Cheng Lv, and K. K. Tan), which appeared in IEEE Trans. Neural Networks, vol. 16, no. 6, pp. 1318–1328, Nov. 2005. ©[2005] IEEE.

obtained. An invariant set and an ultimate bound are derived. A mathematic proof is given to prove the convergence. Simulation results and discussions will be provided in Section 2.3.4. Conclusions are drawn in Section 2.3.5.

The third part of the chapter compares the convergent properties of Oja's algorithm with Xu's LMSER's. The chapter concludes in the Section 2.5.

2.2   Preliminaries

Let $C = E\left[x(k)x^T(k)\right]$. Suppose that λ_p is the smallest nonzero eigenvalue; that is, $\lambda_p > 0$ but $\lambda_j = 0 (j = p+1, \ldots, n)$. Denote by σ, the largest eigenvalue of C. Suppose that the multiplicity of σ is $m (1 \le m \le p \le n)$, then,

$$\sigma = \lambda_1 = \ldots = \lambda_m.$$

Suppose that $\{v_i | i = 1, \ldots, n\}$ is an orthonormal basis in R^n such that each v_i is a unit eigenvector of C associated with the eigenvalue λ_i. Denote by V_σ the eigensubspace of the largest eigenvalue σ; that is,

$$V_\sigma = span\{v_1, \ldots, v_m\}.$$

Denoting by $V_\sigma^\perp$ the subspace, which is perpendicular to V_σ, clearly,

$$V_\sigma^\perp = span\{v_{m+1}, \ldots, v_n\}.$$

The following lemma will be useful in this chapter.

Lemma 2.1 *It holds that*

$$[1 + \eta(\sigma - s)]^2 s \le \frac{4}{27\eta} \cdot (1 + \eta\sigma)^3$$

for all $s \in [0, \sigma + 1/\eta]$.

Proof: Define a differentiable function

$$f(s) = [1 + \eta(\sigma - s)]^2 s$$

for $0 \le s \le \sigma + 1/\eta$. It follows that

$$\dot{f}(s) = [1 + \eta(\sigma - s)] \cdot [1 + \eta(\sigma - 3s)]$$

for $0 \le s \le \sigma + 1/\eta$. Denote

$$\xi = \frac{1}{3}\left(\sigma + \frac{1}{\eta}\right).$$

Then,

$$\dot{f}(s) \begin{cases} > 0, & \text{if } 0 \le s \le \xi \\ = 0, & \text{if } s = \xi \\ < 0, & \text{if } \xi \le s < \sigma + 1/\eta \\ = 0, & \text{if } s = \sigma + 1/\eta. \end{cases}$$

This shows that ξ must be the maximum point of the the function $f(s)$ on the interval $[0, \sigma + 1/\eta]$. Then, it holds that,

$$f(s) \le f(\xi) = \frac{4}{27\eta} \cdot (1 + \eta\sigma)^3,$$

for all $0 \le s \le \sigma + 1/\eta$. The proof is completed.

2.3 Oja's Algorithm

In 1982, Oja used a simple linear single neural network in Figure 1.1 to extract the first principal component from the input data [134]. Based on the well-known Hebbian learning rule, Oja proposed a principal component analysis (PCA) learning algorithm to update the weights of the network. This network, under Oja's algorithm, is able to extract a principal component from input data adaptively. The results have been useful in online data processing applications.

In [210], Zufiria proposed to study the dynamical behaviour of Oja's learning algorithm by analyzing a related DDT algorithm. Certain local and global dynamical properties of DDT algorithm have been reported in [210]. Zufiria has shown that some trajectories of the DDT algorithm can be divergent and even chaotic in nature. Certain local stability results have also been reported. Following the DDT algorithm approach of [210], a generalized Oja's PCA learning algorithm, called bounded PCA algorithm, has been studied in [202].

An important problem to now address – Is the corresponding DDT algorithm of Oja's algorithm in [210] convergent? Since the work of [210], clearly, the DDT algorithm has not been shown to be convergent globally. Moreover, some trajectories may diverge and even become chaotic. Thus, it is quite interesting to explore whether there exists invariant sets to retain the trajectories within the sets and based on these invariant sets to explore the bigger issue of convergence. This section will study these problems in details. The DDT algorithm has many equilibria and thus the study of convergence belongs to a multistability problem. Multistability analysis has recently received attractive attentions, see, for examples, [194], [191], [192], [193]. Novel and deep results regarding convergence of the DDT algorithm will be discussed in this section. A large number of invariant sets will be derived. Any trajectory of the DDT algorithm starting from an invariant set will remain in it indefinitely. This is interesting because it provides a method to avoid divergence of the DDT algorithm by selecting a proper initial vector of the DDT algorithm from the invariant set. Moreover, the convergence of the DDT algorithm can now be proved. In the next section, we will show that almost all trajectories of the DDT algorithm starting from an invariant set will converge to an eigenvector associated with the largest eigenvalue of the correlations matrix. Consequently, it confirms that one can use a constant learning rate in Oja's PCA learning algorithm to guarantee convergence if initial vectors are suitably chosen.

2.3.1   Oja's Algorithm and the DCT Method

With the network (1.1) and input sequence (1.2), the basic Oja's PCA learning algorithm for weight updating is described as

$$w(k+1) = w(k) + \eta y(k) \left[x(k) - y(k)w(k) \right], \tag{2.1}$$

for $k \geq 0$, where $\eta(k) > 0$ is the learning rate.

It is very difficult to study the convergence of (2.1) directly. To indirectly interpret the convergence of (2.1), traditionally, the DCT method is used. This method can be simply described as follows: applying a fundamental stochastic approximation theorem [117] to (2.1), a deterministic continuous time systems can be obtained:

$$\frac{dw(t)}{dt} = Cw(t) - w^T(t)Cw(t) \cdot w(t), \quad t \geq 0,$$

where $C = E\left[x(k)x^T(k)\right]$ denotes the covariance matrix of

$$\{x(k) \, | \, x(k) \in R^n (k = 0, 1, 2, \ldots)\} .$$

The convergence of this DCT system is then used to interpret the convergence of (2.1). To transform (2.1) into the above DCT system, the following restrictive conditions are required:

(C1): $\sum_{k=1}^{+\infty} \eta^2(k) < +\infty$;

(C2): $\sup_k E\left\{ |y(k)\left[x(k) - y(k)w(k)\right]|^2 \right\} < +\infty$.

By the condition (C1), clearly, it implies that $\eta(k) \to 0$ as $k \to +\infty$. However, in many practical applications, $\eta(k)$ is often set to a small constant due to the round-off limitation and speed requirement [210], [201]. The condition (C2) is also difficult to be satisfied [210]. Thus, these conditions are unrealistic in practical applications, and the previous DCT analysis is not directly applicable [210], [201].

2.3.2   DDT Formulation

To overcome the problem of learning rate converging toward zero, a DDT method is proposed in [210]. Unlike the DCT method to transform (2.1) into a continuous time deterministic system, the DDT method transform

(2.1) into a discrete time deterministic algorithm. The DDT algorithm can be formulated as follows. Applying the conditional expectation operator $E\{w(k+1)/w(0), x(i), i < n\}$ to (2.1) and identifying the conditional expected value as the next iterate in the system [210], a DDT system can be obtained and given as

$$w(k+1) = w(k) + \eta \left[Cw(k) - w^T(k)Cw(k)w(k) \right], \qquad (2.2)$$

for $k \geq 0$, where $C = E\left[x(k)x^T(k)\right]$ is the covariance matrix. The DDT method has some obvious advantages. First, it preserves the discrete time nature as that of the original learning algorithm. Second, it is not necessary to satisfy some unrealistic conditions of the DCT approach. Thus, it allows the learning rate to be constant. The main purpose of this section is to study the convergence of (2.2) subject to the learning rate η being some constant.

Clearly, from (2.2), (1.7), (1.8), it holds that

$$z_i(k+1) = \left[1 + \eta \left(\lambda_i - w^T(k)Cw(k) \right) \right] z_i(k), (i = 1, \ldots, n), \qquad (2.3)$$

for $k \geq 0$.

Definition 2.1 *A point $w^* \in R^n$ is called an equilibrium of (2.2), if and only if*

$$w^* = w^* + \eta \left[Cw^* - (w^*)^T Cw^* w^* \right].$$

Clearly, the set of all equilibrium points of (2.2) is

$$\{v_1, \ldots, v_n\} \cup \{-v_1, \ldots, -v_n\} \cup \{0\}.$$

2.3.3   Invariant Sets and Convergence Results

Some trajectories of (2.2) may diverge. Consider a simple example of a one dimensional equation,

$$w(k+1) = w(k) + \eta \left[1 - w^2(k) \right] w(k), k \geq 0.$$

If $w(k) \geq 1 + 1/\eta$, it holds that

$$\eta w^2(k) - w(k) - \eta - 1 \geq 0.$$

Then,

$$\begin{aligned} |w(k+1)| &= w^2(k) \cdot \frac{\eta w^2(k) - \eta - 1}{w(k)} \\ &\geq w^2(k). \end{aligned}$$

Clearly, the trajectory approaches toward infinity if $w(0) \geq 1+1/\eta$. Other examples about divergence of Oja's algorithm can be found in [210]. A problem to address is therefore to find out the conditions under which the trajectories can be bounded. The results forthcoming will be very important for applications. Next, we will prove an interesting theorem, which gives an invariant set of (2.2). The trajectories, thus, can be guaranteed to be bounded.

Theorem 2.1 *Denote*

$$S = \left\{ w \;\middle|\; w \in R^n, w^T C w \le \frac{1}{\eta} \right\}.$$

If

$$\eta\sigma \le \frac{3\sqrt[3]{2} - 2}{2} \approx 0.8899,$$

then S is an invariant set of (2.2).

Proof: Given any $k \ge 0$, suppose $w(k) \in S$, that is,

$$0 \le w^T(k) C w(k) \le \frac{1}{\eta}.$$

Then,

$$1 + \eta\left(\lambda_i - w^T(k)Cw(k)\right) \ge \eta\lambda_i \ge 0, (i = 1, \ldots, n).$$

From (2.2), (1.7), (1.8), and (2.3), it follows that

$$
\begin{aligned}
&w^T(k+1)Cw(k+1) \\
=\ & \sum_{i=1}^{n} \lambda_i z_i^2(k+1) \\
=\ & \sum_{i=1}^{n} \lambda_i \left[1 + \eta\left(\lambda_i - w^T(k)Cw(k)\right)\right]^2 z_i^2(k) \\
\le\ & \sum_{i=1}^{n} \lambda_i \left[1 + \eta\left(\sigma - w^T(k)Cw(k)\right)\right]^2 z_i^2(k) \\
=\ & \left[1 + \eta\left(\sigma - w^T(k)Cw(k)\right)\right]^2 \cdot w^T(k)Cw(k) \\
\le\ & \max_{0 \le s \le 1/\eta} \left\{ \left[1 + \eta\left(\sigma - s\right)\right]^2 \cdot s \right\} \\
\le\ & \max_{0 \le s \le \sigma + 1/\eta} \left\{ \left[1 + \eta\left(\sigma - s\right)\right]^2 \cdot s \right\}.
\end{aligned}
$$

Using Lemma 2.1,

$$w^T(k+1)Cw(k+1) \le \frac{4}{27\eta}\left(1 + \eta\sigma\right)^3.$$

By

$$\eta\sigma \le \frac{3\sqrt[3]{2} - 2}{2},$$

it follows that

$$\frac{4}{27\eta}\left(1 + \eta\sigma\right)^3 \le \frac{1}{\eta}.$$

Thus,

$$w^T(k+1)Cw(k+1) \leq \frac{1}{\eta}.$$

This shows that $w(k+1) \in S$, that is, S is an invariant set. The proof is completed. Next, we will provide a deeper result, which will give a family of invariant sets.

Theorem 2.2 *Given any constant l such that $1 \leq l \leq 2$, if*

$$\eta\sigma \leq \frac{3\sqrt[3]{2l} - 2}{2},$$

then, the set

$$S(l) = \left\{ w \,\middle|\, w \in R^n, w^T Cw \leq \lambda_p + \frac{l}{\eta} \right\}$$

is an invariant set of (2.2).

Proof: Given any $k \geq 0$, suppose $w(k) \in S(l)$, we will prove that $w(k+1) \in S(l)$. Clearly,

$$0 \leq w^T(k)Cw(k) \leq \lambda_p + \frac{l}{\eta}.$$

If

$$\sigma \leq w^T(k)Cw(k) \leq \lambda_p + \frac{l}{\eta},$$

then

$$\left[1 + \eta\left(\lambda_i - w^T(k)Cw(k)\right)\right]^2$$
$$= 1 + \eta^2\left(w^T(k)Cw(k) - \lambda_i\right)\left(w^T(k)Cw(k) - \lambda_i - \frac{2}{\eta}\right)$$
$$\leq 1, (i = 1, \dots, p).$$

Thus,

$$w^T(k+1)Cw(k+1)$$
$$= \sum_{i=1}^{n} \lambda_i z_i^2(k+1)$$
$$= \sum_{i=1}^{p} \lambda_i z_i^2(k+1)$$
$$= \sum_{i=1}^{p} \lambda_i \left[1 + \eta\left(\lambda_i - w^T(k)Cw(k)\right)\right]^2 z_i^2(k)$$
$$\leq \sum_{i=1}^{p} \lambda_i z_i^2(k)$$
$$= w^T(k)Cw(k)$$
$$\leq \lambda_p + \frac{l}{\eta}.$$

If
$$0 \leq w^T(k)Cw(k) \leq \sigma,$$

then it can be checked that

$$
\begin{aligned}
& \left[1 + \eta\left(\lambda_i - w^T(k)Cw(k)\right)\right]^2 \\
\leq\ & \left[1 + \eta\left(\sigma - w^T(k)Cw(k)\right)\right]^2,
\end{aligned}
$$

for $i = 1, \ldots, n$. By Lemma 1,

$$
\begin{aligned}
& w^T(k+1)Cw(k+1) \\
=\ & \sum_{i=1}^{n} \lambda_i \left[1 + \eta\left(\lambda_i - w^T(k)Cw(k)\right)\right]^2 z_i^2(k) \\
\leq\ & \left[1 + \eta\left(\sigma - w^T(k)Cw(k)\right)\right]^2 \cdot w^T(k)Cw(k) \\
\leq\ & \max_{0 \leq s \leq \sigma} \left\{ \left[1 + \eta\left(\sigma - s\right)\right]^2 s \right\} \\
\leq\ & \max_{0 \leq s \leq \sigma + 1/\eta} \left\{ \left[1 + \eta\left(\sigma - s\right)\right]^2 s \right\} \\
\leq\ & \frac{4}{27\eta}(1 + \eta\sigma)^3 \\
\leq\ & \frac{l}{\eta} \\
\leq\ & \lambda_p + \frac{l}{\eta}.
\end{aligned}
$$

This shows that $w(k+1) \in S(l)$. The proof is completed.

Some invariant sets will follow from the above analysis. The nondivergence can thus be guaranteed by selecting initial vectors from the invariant sets. Next, we will further establish convergence results.

From (1.7), for each $k \geq 0$, $w(k)$ can be represented as

$$w(k) = \sum_{i=1}^{m} z_i(k)v_i + \sum_{j=m+1}^{n} z_j(k)v_j,$$

where $z_i(k)(i = 1, \ldots, n)$ are some constants for each k. In particular,

$$w(0) = \sum_{i=1}^{m} z_i(0)v_i + \sum_{j=m+1}^{n} z_j(0)v_j.$$

Clearly, the convergence of $w(k)$ can be determined by the convergence of $z_i(k)(i = 1, \ldots, n)$. Next, we analyze the convergence of $w(k)$ by studying the convergence of $z_i(k)(i = 1, \ldots, m)$ and $z_i(k)(i = m+1, \ldots, n)$, respectively.

Lemma 2.2 *Suppose that*

$$\eta\sigma \leq \frac{3\sqrt[3]{2} - 2}{2} \approx 0.8899,$$

if $w(0) \in S$ and $w(0) \notin V_\sigma^\perp$, then there exist constants $\theta_1 > 0$ and $\Pi_1 \geq 0$ such that

$$\sum_{j=m+1}^{n} z_j^2(k) \leq \Pi_1 \cdot e^{-\theta_1 k},$$

for all $k \geq 0$, where

$$\theta_1 = \ln \left(\frac{1 + \eta\sigma}{1 + \eta\lambda_{m+1}} \right)^2 > 0.$$

Proof : Since $w(0) \notin V_\sigma^\perp$, there must exist some $i(1 \leq i \leq m)$ such that $z_i(0) \neq 0$. Without loss of generality, assume that $z_1(0) \neq 0$.

From (2.3), it follows that

$$z_i(k + 1) = [1 + \eta(\sigma - w^T(k)Cw(k))]z_i(k), 1 \leq i \leq m, \qquad (2.4)$$

and

$$z_j(k + 1) = [1 + \eta(\lambda_j - w^T(k)Cw(k))]z_j(k), m + 1 \leq j \leq n, \qquad (2.5)$$

for $k \geq 0$.

By Lemma 2.1, S is an invariant set, then $w(k) \in S$ for all $k \geq 0$. Given any $i(1 \leq i \leq n)$, it holds that

$$\begin{aligned}
1 + \eta\left(\lambda_i - w^T(k)Cw(k)\right) &= \eta\lambda_i + \left(1 - \eta w^T(k)Cw(k)\right) \\
&\geq \eta\lambda_i > 0,
\end{aligned}$$

for $k \geq 1$. Then, from (2.4) and (2.5), for each $j(m + 1 \leq j \leq n)$, it follows that

$$\begin{aligned}
\left[\frac{z_j(k+1)}{z_1(k+1)} \right]^2 &= \left[\frac{1 + \eta(\lambda_j - w^T(k)Cw(k))}{1 + \eta(\sigma - w^T(k)Cw(k))} \right]^2 \cdot \left[\frac{z_j(k)}{z_1(k)} \right]^2 \\
&\leq \left[1 - \frac{\eta(\sigma - \lambda_j)}{1 + \eta(\sigma - w^T(k)Cw(k))} \right]^2 \cdot \left[\frac{z_j(k)}{z_1(k)} \right]^2 \\
&\leq \left(\frac{1 + \eta\lambda_j}{1 + \eta\sigma} \right)^2 \cdot \left[\frac{z_j(k)}{z_1(k)} \right]^2 \\
&\leq \left(\frac{1 + \eta\lambda_{m+1}}{1 + \eta\sigma} \right)^{2(k+1)} \cdot \left[\frac{z_j(0)}{z_1(0)} \right]^2 \\
&= \left[\frac{z_j(0)}{z_1(0)} \right]^2 \cdot e^{-\theta_1(k+1)},
\end{aligned}$$

for all $k \geq 1$.

By the invariance of S, $z_1(k)$ must be bounded; that is, there exists a constant $d > 0$ such that $z_1^2(k) \leq d$ for all $k \geq 0$. Then,

$$\sum_{j=m+1}^{n} z_j^2(k) = \sum_{j=m+1}^{n} \left[\frac{z_j(k)}{z_1(k)}\right]^2 \cdot z_1^2(k) \leq \Pi_1 e^{-\theta_1 k},$$

for $k \geq 0$, where

$$\Pi_1 = d \sum_{j=m+1}^{n} \left[\frac{z_j(0)}{z_1(0)}\right]^2 \geq 0.$$

This completes the proof.

Lemma 2.3 *Suppose that*

$$\eta \sigma \leq \frac{3\sqrt[3]{2} - 2}{2} \approx 0.8899.$$

If $w(0) \in S$ and $w(0) \neq 0$, then

$$\eta w^T(k+1)Cw(k+1) \geq \min\left\{(\eta\lambda_p)^3, \eta w^T(0)Cw(0)\right\}$$

for all $k \geq 0$.

Proof: Since $w(0) \in S$, by Theorem 1, $w(k) \in S$ for all $k \geq 0$, that is, $w^T(k)Cw(k) \leq 1/\eta$, then

$$1 + \eta\left(\lambda_i - w^T(k)Cw(k)\right) \geq \eta\lambda_i \geq \eta\lambda_p > 0, (i = 1, \ldots, p),$$

for all $k \geq 0$.

If $\lambda_p \leq w^T(k)Cw(k) \leq 1/\eta$, from (1.8) and (2.3), it follows that

$$\eta w^T(k+1)Cw(k+1)$$

$$= \eta \sum_{i=1}^{n} \lambda_i z_i^2(k+1)$$

$$= \eta \sum_{i=1}^{p} \lambda_i z_i^2(k+1)$$

$$= \eta \sum_{i=1}^{p} \left[1 + \eta\left(\lambda_i - w^T(k)Cw(k)\right)\right]^2 \lambda_i z_i^2(k)$$

$$\geq \left[\eta\lambda_p + \left(1 - \eta w^T(k)Cw(k)\right)\right]^2 \cdot \eta w^T(k)Cw(k)$$

$$\geq (\eta\lambda_p)^2 \cdot \eta w^T(k)Cw(k)$$

$$\geq (\eta\lambda_p)^3 \tag{2.6}$$

for $k \geq 0$.

If $w^T(k)Cw(k) \leq \lambda_p$, then,

$$w^T(k+1)Cw(k+1)$$
$$= \sum_{i=1}^{p} \left[1 + \eta\left(\lambda_i - w^T(k)Cw(k)\right)\right]^2 \lambda_i z_i^2(k)$$
$$\geq \left[1 + \eta\left(\lambda_p - w^T(k)Cw(k)\right)\right]^2 w^T(k)Cw(k)$$
$$\geq w^T(k)Cw(k) \tag{2.7}$$

for $k \geq 0$.

It follows from (2.6) and (2.7) that

$$\eta w^T(k+1)Cw(k+1) \geq \min\left\{(\eta\lambda_p)^3, \eta w^T(k)Cw(k)\right\}$$
$$\geq \min\left\{(\eta\lambda_p)^3, \eta w^T(0)Cw(0)\right\}$$

for all $k \geq 0$. This completes the proof.

The above lemma shows that if $w(0) \neq 0$, the trajectory starting from $w(0)$ is lower bounded, then it will never converge to zero. Clearly, the zero vector must be an unstable equilibrium of (2.2).

Lemma 2.4 *Suppose that*

$$\eta\sigma \leq \frac{3\sqrt[3]{2} - 2}{2} \approx 0.8899,$$

and $w(0) \in S$, $w(0) \notin V_\sigma$. Then, there exists constants $\theta_2 > 0$ and $\Pi_2 > 0$ such that

$$\left|\sigma - w^T(k+1)Cw(k+1)\right|$$
$$\leq k \cdot \Pi_2 \cdot \left[e^{-\theta_2(k+1)} + \max\left\{e^{-\theta_2 k}, e^{-\theta_1 k}\right\}\right]$$

for all $k \geq 0$, where

$$\begin{cases} \theta_2 = -\ln\delta, \\ \delta = \max\left\{(1-\gamma)^2, \eta\sigma\right\}, \\ \gamma = \min\left\{(\eta\lambda_p)^3, \eta w^T(0)Cw(0)\right\}, \end{cases}$$

and $0 < \delta < 1$, $0 < \gamma < 1$.

Proof: Clearly,

$$w^T(k+1)Cw(k+1)$$

$$= \sum_{i=1}^{n} \lambda_i \left[1 + \eta \left(\lambda_i - w^T(k)Cw(k)\right)\right]^2 z_i^2(k)$$

$$= \sum_{i=1}^{n} \lambda_i \left[1 + \eta \left(\sigma - w^T(k)Cw(k)\right)\right]^2 z_i^2(k)$$

$$+ \sum_{i=m+1}^{n} \lambda_i \left[1 + \eta \left(\lambda_i - w^T(k)Cw(k)\right)\right]^2 z_i^2(k)$$

$$- \sum_{i=m+1}^{n} \lambda_i \left[1 + \eta \left(\sigma - w^T(k)Cw(k)\right)\right]^2 z_i^2(k)$$

$$= \left[1 + \eta \left(\sigma - w^T(k)Cw(k)\right)\right]^2 \cdot w^T(k)Cw(k) + H(k)$$

for $k \geq 0$, where

$$H(k) = \eta \sum_{i=m+1}^{n} \lambda_i (\lambda_i - \sigma)$$

$$\times \left[2 + \eta \left(\lambda_i + \sigma - 2w^T(k)Cw(k)\right)\right] z_i^2(k).$$

Then,

$$\sigma - w^T(k+1)Cw(k+1)$$

$$= \left[\sigma - w^T(k)Cw(k)\right]$$

$$\times \left[\left(1 - \eta w^T(k)Cw(k)\right)^2 - \eta^2 \sigma w^T(k)Cw(k)\right]$$

$$-H(k),$$

for $k \geq 0$. Denote

$$V(k) = \left|\sigma - w^T(k)Cw(k)\right|,$$

for $k \geq 0$. By the invariance of S, $w(k) \in S$ for $k \geq 0$. It follows that

$$V(k+1)$$

$$\leq V(k) \cdot \left|\left(1 - \eta w^T(k)Cw(k)\right)^2 - \eta^2 \sigma w^T(k)Cw(k)\right|$$

$$+|H(k)|$$

$$\leq \max\left\{\left(1 - \eta w^T(k)Cw(k)\right)^2, \eta^2 \sigma w^T(k)Cw(k)\right\} \cdot V(k)$$

$$+|H(k)|$$

$$\leq \max\left\{\left(1 - \eta w^T(k)Cw(k)\right)^2, \eta\sigma\right\} \cdot V(k) + |H(k)|.$$

Denote

$$\gamma = \min\left\{(\eta\lambda_p)^3, \eta w^T(0)Cw(0)\right\}.$$

Clearly, $0 < \gamma < 1$. By Lemma 2.3, it holds that

$$\gamma \leq \eta w^T(k)Cw(k) \leq 1.$$

Denote

$$\delta = \max\left\{(1-\gamma)^2, \eta\sigma\right\}.$$

Clearly, $0 < \delta < 1$. Then,

$$V(k+1) \leq \delta \cdot V(k) + |H(k)|, k \geq 0.$$

By Lemma 2.2,

$$
\begin{aligned}
&|H(k)| \\
\leq\ & \eta \sum_{j=m+1}^{n} \lambda_j (\sigma - \lambda_j)\left[\eta(\sigma + \lambda_j) + 2\right] z_j^2(k) \\
\leq\ & \eta \sum_{j=m+1}^{n} \left[\eta\lambda_j(\sigma^2 - \lambda_j^2) + 2\lambda_j(\sigma - \lambda_j)\right] z_j^2(k) \\
\leq\ & \eta \sum_{j=m+1}^{n} \left(\eta\lambda_j\sigma^2 + 2\lambda_j\sigma\right) z_j^2(k) \\
\leq\ & \sigma(\eta\sigma)(\eta\sigma + 2) \sum_{j=m+1}^{n} z_j^2(k) \\
\leq\ & 3\sigma \cdot \sum_{j=m+1}^{n} z_j^2(k) \\
\leq\ & 3\sigma\Pi_1 \cdot e^{-\theta_1 k}
\end{aligned}
$$

for all $k \geq 0$. Then,

$$
\begin{aligned}
V(k+1) &\leq \delta^{k+1}V(0) + 3\sigma\Pi_1 \cdot \sum_{r=0}^{k} \left(\delta e^{\theta_1}\right)^r e^{-\theta_1 k} \\
&\leq \delta^{k+1}V(0) + 3k\sigma\Pi_1 \cdot \max\left\{\delta^k, e^{-\theta_1 k}\right\} \\
&\leq k \cdot \Pi_2 \cdot \left[e^{-\theta_2(k+1)} + \max\left\{e^{-\theta_2 k}, e^{-\theta_1 k}\right\}\right],
\end{aligned}
$$

where $\theta_2 = -\ln\delta > 0$ and

$$\Pi_2 = \max\left\{\left|\sigma - w^T(0)Cw(0)\right|, 3\sigma\Pi_1\right\} > 0.$$

The proof is completed.

Lemma 2.5 *Suppose there exists constants $\theta > 0$ and $\Pi > 0$ such that*

$$\eta\left|\left(\sigma - w^T(k)Cw(k)\right)z_i(k)\right| \leq k \cdot \Pi e^{-\theta k}, (i = 1, \ldots, m)$$

for $k \geq 0$. *Then,*

$$\lim_{k \to +\infty} z_i(k) = z_i^*, (i = 1, \ldots, m),$$

where $z_i^* (i = 1, \ldots, m)$ *are constants.*

Proof: Given any $\epsilon > 0$, there exists a $K \geq 1$ such that

$$\frac{\Pi K e^{-\theta K}}{(1 - e^{-\theta})^2} \leq \epsilon.$$

For any $k_1 > k_2 \geq K$, it follows that

$$
\begin{aligned}
|z_i(k_1) - z_i(k_2)| &= \left| \sum_{r=k_2}^{k_1-1} [z_i(r+1) - z_i(r)] \right| \\
&\leq \eta \sum_{r=k_2}^{k_1-1} \left| \left(\sigma - w^T(r) C w(r) \right) z_i(r) \right| \\
&\leq \Pi \sum_{r=k_2}^{k_1-1} r e^{-\theta r} \\
&\leq \Pi \sum_{r=K}^{+\infty} r e^{-\theta r} \\
&\leq \Pi K e^{-\theta K} \cdot \sum_{r=0}^{+\infty} r \left(e^{-\theta} \right)^{r-1} \\
&= \frac{\Pi K e^{-\theta K}}{(1 - e^{-\theta})^2} \\
&\leq \epsilon, \quad (i = 1, \ldots, m).
\end{aligned}
$$

This shows that each sequence $\{z_i(k)\}$ is a *Cauchy sequence*. By *Cauchy Convergence Principle*, there must exist constants $z_i^* (i = 1, \ldots, m)$ such that

$$\lim_{k \to +\infty} z_i(k) = z_i^*, \quad (i = 1, \ldots, m).$$

This completes the proof.

Theorem 2.3 *Suppose that*

$$\eta \sigma \leq \frac{3 \sqrt[3]{2} - 2}{2} \approx 0.8899,$$

if $w(0)$ *and* $w(0) \notin V_\sigma^\perp$, *then the trajectory of (2.2) starting from* $w(0)$ *will converge to a unit eigenvector associated with the largest eigenvalue of the covariance matrix* C.

Proof: By Lemma 2.2, there exists constants $\theta_1 > 0$ and $\Pi_1 \geq 0$ such that

$$\sum_{j=m+1}^{n} z_j^2(k) \leq \Pi_1 e^{-\theta_1 k},$$

for all $k \geq 0$. By Lemma 2.4, there exists constants $\theta_2 > 0$ and $\Pi_2 > 0$ such that

$$\left| \sigma - w^T(k+1)Cw(k+1) \right|$$
$$\leq k \cdot \Pi_2 \cdot \left[e^{-\theta_2(k+1)} + \max\left\{ e^{-\theta_2 k}, e^{-\theta_1 k} \right\} \right]$$

for all $k \geq 0$.

Obviously, there exists constants $\theta > 0$ and $\Pi > 0$ such that

$$\eta \left| \left(\sigma - w^T(k)Cw(k) \right) z_i(k) \right| \leq k \cdot \Pi e^{-\theta k}, (i = 1, \ldots, m)$$

for $k \geq 0$.

Using Lemma 2.5 and Lemma 2.2, it follows that

$$\begin{cases} \lim_{t \to +\infty} z_i(k) = z_i^*, (i = 1, \ldots, m) \\ \lim_{t \to +\infty} z_i(k) = 0, (i = m+1, \ldots, n). \end{cases}$$

Then,

$$\lim_{t \to +\infty} w(k) = \sum_{i=1}^{m} z_i^* v_i \in V_\sigma.$$

It is easy to see that

$$\sum_{i=1}^{m} (z_i^*)^2 = 1.$$

The proof is completed.

In the above theorem, it requires that the initial $w(0) \notin V_\sigma^\perp$ so that there is at least one $z_i(0) \neq 0 (1 \leq i \leq m)$. Since the dimension of $V_\sigma^\perp$ is less than that of R^n, the measure of $V_\sigma^\perp$ is zero, the condition $w(0) \notin V_\sigma^\perp$ is easy to meet in practical applications. In fact, any small disturbance can result in $w(0) \notin V_\sigma^\perp$. From this perspective, we can deduce that almost all trajectories starting from S will converge to a unit eigenvector associated with the largest eigenvalue of the covariance matrix of C. Also, it is worthwhile to point out that $w(0) \notin V_\sigma^\perp$ implies $w(0) \neq 0$.

2.3.4 Simulation and Discussions

2.3.4.1 Illustration of Invariant Sets

Theorem 2.1 gives an invariant set S, while Theorem 2.2 gives a family of invariant sets $S(l)(1 \leq l \leq 2)$. Clearly, $S \subset S(l)$. Let us first illustrate the invariance of $S(2)$, the largest invariant set in Theorem 2.2. Consider a simple example of (2.2) with two dimension, where

$$C = \left[\begin{array}{cc} 1 & 0 \\ 0 & 0.2 \end{array} \right].$$

It is easy to see that

$$S(2) = \left\{ \left[\begin{array}{c} w_1 \\ w_2 \end{array} \right] \middle| w_1^2 + 0.2w_2^2 \le 1.6481 \right\}.$$

In Figure 2.1, the ellipsoid contained part is $S(2)$. Figure 2.1 shows that 80 trajectories of (2.2) starting from points in $S(2)$ are all contained in $S(2)$ for 5000 iterations with $\eta\sigma = 1.3811$. It clearly shows the invariance of $S(2)$.

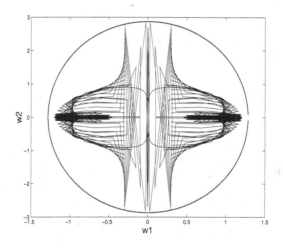

FIGURE 2.1
Invariance of $S(2)$. Eighty trajectories starting from points in $S(2)$ remain in $S(2)$.

Though the set $S(2)$ is invariant, it does not imply that each trajectory in $S(2)$ is convergent. In fact, some trajectories contained in $S(2)$ may not converge. Figure 2.2 shows such a nonconvergent trajectory starting from $w(0) = [0.2, 2.5]^T$ with $\eta\sigma = 1.3811$.

2.3.4.2 Selection of Initial Vectors

Theorem 2.3 shows that trajectories starting from $S - \{0\}$ will converge to a unit eigenvector associated with the largest eigenvalue. Thus, choosing initial vectors from S is essential for convergence. However, this selection needs information of the matrix C. Generally, it is not practical. Next, a more practical domain for initial vector selection is suggested.

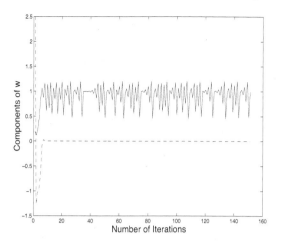

FIGURE 2.2
A nonconvergent trajectory in $S(2)$.

Denote a unit hyper-sphere in R^n by

$$U = \left\{ w \,\middle|\, w^T w \le 1 \right\}.$$

For any $w \in U$, it holds that

$$w^T C w \le \sigma w^T w \le \sigma \le \frac{1}{\eta},$$

i.e., $w \in S$. Thus, $U \subset S$. This suggests that one can select initial vector w in $U - \{0\}$. From the application point of view, choosing initial vectors from $U - \{0\}$ is an easy task. For example, we may simply generate an initial vector in U by

$$w(0) = \frac{1}{\sqrt{n}} \cdot random(n).$$

2.3.4.3   Selection of Learning Rate

From the analysis in the last section, to guarantee convergence, the learning rate should satisfy $0 < \eta\sigma \le 0.8899$. If we choose $\eta\sigma$ close to 0, from Lemma 2.2, it may slow down the convergence of $z_i(k)(j = m+1, \ldots, n)$. If we choose $\eta\sigma$ close to 0.8899, from Lemma 2.4, it may slow down the convergence of $z_i(k)(i = 1, \ldots, m)$. To get a balance to suitably set $\eta\sigma$ for fast convergence, it would be better if

$$\theta_1 \approx \theta_2.$$

Under some mild conditions, it holds that

$$\theta_1 = \ln\left(\frac{1 + \eta\sigma}{1 + \eta\lambda_{m+1}}\right) \approx \ln\left(1 + \eta\sigma\right), \theta_2 \approx \ln\left(\frac{1}{\eta\sigma}\right).$$

Solving

$$\ln\left(1 + \eta\sigma\right) \approx \ln\left(\frac{1}{\eta\sigma}\right),$$

it follows that

$$\eta\sigma \approx 0.618.$$

This shows that one can set $\eta\sigma$ around 0.618 to achieve fast convergence. The next simulations will support this.

We randomly generated a 5×5 symmetric nonnegative definite matrix as

$$C = \begin{bmatrix} 0.1947 & 0.3349 & 0.2783 & 0.1494 & 0.0364 \\ 0.3349 & 0.8052 & 0.5105 & 0.1716 & 0.0361 \\ 0.2783 & 0.5105 & 0.4149 & 0.1916 & 0.0348 \\ 0.1494 & 0.1716 & 0.1916 & 0.1608 & 0.0518 \\ 0.0364 & 0.0361 & 0.0348 & 0.0518 & 0.0262 \end{bmatrix}.$$

Selecting an initial vector randomly in $U - \{0\}$ as

$$w(0) = \begin{bmatrix} 0.0706, 0.2185, 0.0776, 0.3105, 0.4214 \end{bmatrix}^T.$$

Figure 2.3 shows the trajectory of (2.2) starting from $w(0)$ converging to an eigenvector

$$w^* = \begin{bmatrix} 0.3574, 0.7322, 0.5299, 0.2296, 0.0510 \end{bmatrix}^T$$

in 17 iterations with precision $\epsilon = 0.00001$ and $\eta\sigma = 0.62$. Clearly, w^* is a unit eigenvector associated with the largest eigenvalue of C.

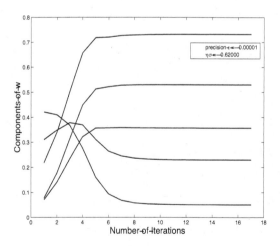

FIGURE 2.3
Convergence of (2.2).

To further illustrate that by setting $\eta\sigma$ around 0.618, the DDT algorithm

(2.2) can really converge fast, we randomly generated three symmetric non-negative definite matrices with dimensions of 5×5, 50×50, and 500×500, respectively. Figure 2.4 shows the performance of (2.2) with iteration step variations corresponding to the value of $\eta\sigma$ from 0.2 to 0.8899. It clearly shows

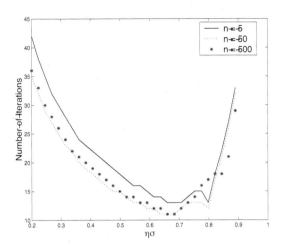

FIGURE 2.4
Iteration step variations with respect to $\eta\sigma$.

that the algorithm (2.2) converges fast if the value of $\eta\sigma$ is selected between 0.6 and 0.7. The simulation in Figure 2.4 also shows that the convergence of (2.2) does not appear to be sensitive to the dimension of (2.2) because the three curves are very close. Besides the above simulation results, we further and extensively simulated the convergence of (2.2) with high dimensions. The ensuing simulation results persist to show similarly satisfactory results.

The above simulations further confirm our important result: the learning rate can assume some constant to guarantee convergence providing the initial weight vector selected from an invariant set. Moreover, the above simulations also show that setting $\eta\sigma$ around 0.618 can expedite the convergence of (2.2). To calculate η, we need to estimate σ, the largest eigenvalue of C. In some applications, based on the problem-specific knowledge, a upper bound of σ can be often estimated [201] without computing the covariance matrix C. However, in some practical applications, especially for online computing, it is difficult to estimate σ. Can we find another method to estimate η? Next, we give a way to estimate η. To make the illustration clearer, let us use the following version [31] of algorithm (2.1) for online computing:

$$w(k+1) = w(k) + \eta(k)\left(C_k w(k) - w^T(k)C_k w(k)w(k)\right) \qquad (2.8)$$

where C_k is a online observation for C [31] and $\lim_{k \to +\infty} C_k = C$. Noting that

$$\lim_{k \to +\infty} w^T(k)C_k w(k) \to \sigma$$

whenever $w(k)$ converges to the unit eigenvector associate with the largest eigenvalue of C, we can simply take $\sigma \approx w^T(k)C_k w(k)$ and then take

$$\eta(k) = \frac{0.618}{w^T(k)C_k w(k)} \qquad (2.9)$$

as an online learning rate of (2.8). When the learning step k becomes sufficiently large, one can trunk $\eta(k)$ to be a constant and take it as a constant learning rate thereafter. Since in (2.9) $\eta(k) \to 0.618/\sigma \neq 0$, it is expected that the convergence could be fast.

Finally, the well-known Lenna picture in Figure 1.5 will be used for illustration. Each online observation C_k (1.10) is used as input for algorithm (2.8). To measure the convergence and accuracy of the algorithm (2.8), we compute the norm of $w(k)$ and the direction cosine at kth update by [31]:

$$\text{Direction Cosine } (k) = \frac{|w^T(k) \cdot \phi|}{\|w(k)\| \cdot \|\phi\|},$$

where ϕ is the true eigenvector associated with the largest eigenvalue of the covariance matrix C computed from all collected samples by the MATLAB 7.0. If both $\|w(k)\|$ and $DirectionCosine(k)$ converge to 1, the algorithm gives the right result. In fact, it clearly shows that $w(k)$ converges to the unit eigenvector associated with the largest eigenvalue of the covariance matrix of C. Figure 2.5 shows the convergence of the direction cosine and the convergence of $\|w(k)\|$. In Figure 2.5, the $\eta(k)$ reaches 0.2527. Since $\eta(k)$ converges to a positive

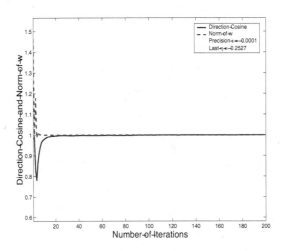

FIGURE 2.5
Convergence of Direction Cosine and norm of w.

constant, the convergent speed is quite fast.

2.3.5 Conclusion

Important results pertaining to the convergence of a DDT algorithm of Oja's PCA learning algorithm have been derived in this section. It shows that almost all trajectories starting from an invariant set converge to the unit eigenvector associated with the largest eigenvalue of the covariance matrix if the learning rate satisfy some simple condition. We suggest using the unit hyper-sphere for conveniently choosing the initial vector to guarantee convergence and the selection of learning rate to achieve fast convergence. Simulation studies are carried out to further illustrate the theory. The most important result is that it confirms that the use of a constant learning rate for Oja's PCA learning algorithm can achieve convergence.

2.4 Xu's LMSER Learning Algorithm

In [184], based on the least mean square error reconstruction, Xu proposed a learning algorithm to perform the true PCA. In this section, by using the DDT method, we will discuss the convergence of Xu's LMSER algorithm, in the one unit case, with a constant learning rate. An invariant set and an evolution ultimate bound are identified. A mathematical proof will be provided to prove the convergence.

2.4.1 Formulation and Preliminaries

Suppose the input sequence (1.2) is a zero mean stationary stochastic process, denoted by $C = E\left[x(k)x^T(k)\right]$, the covariance matrix of the input data set, and let C_k be an online observation of C, then the Xu's learning algorithm can be described by the following stochastic difference equation:

$$
\begin{aligned}
w_l(k+1) =& w_l(k) + \eta(k)[2C_k w_l(k) \\
& - \sum_{i=1}^{l-1} C_k w_l(k) w_i^T(k) w_i(k) - C_k(k) w_l(k) w_l^T w_l(k) \\
& - \sum_{i=1}^{l-1} w_i(k) w_i(k)^T C_k w_l(k) - \left(w_l^T(k) C_k w_l(k)\right) w_l(k)], \quad (2.10)
\end{aligned}
$$

for $\eta(k) > 0$, where w_l is the input weight vector for the lth neuron. Obviously, if $l = 1$, the algorithm is as follows:

$$
\begin{aligned}
w(k+1) =& w(k) + \eta(k)[2C_k w(k) \\
& - C_k(k) w(k) w^T(k) w(k) - \left(w^T(k) C_k w(k)\right) w(k)], \quad (2.11)
\end{aligned}
$$

for $\eta(k) > 0$. The algorithm (2.11) approximates the Oja's one unit rule when the normal of weight vector approaches one [184]. However, it shows some different dynamical behaviors. In this chapter, by studying an associated DDT algorithm, we will discuss the convergence of (2.11), which is a special case of the one unit version of Xu's LMSER algorithm. By taking the conditional expectation $E\{w(k+1)/w(0), w(i), i < k\}$, the DDT algorithm of (2.11) with a constant learning rate is given as follows:

$$
\begin{aligned}
w(k+1) =& w(k) + \eta[2Cw(k) \\
& - C(k)w(k)w^T(k)w(k) - \left(w^T(k)Cw(k)\right)w(k)],
\end{aligned} \qquad (2.12)
$$

for $\eta > 0$ and $C = E[x(k)x^T(k)]$ is the covariance matrix.

From (1.7), (1.8), (2.12), it holds that

$$
z_i(k+1) = \left[1 + \eta\left(\lambda_i(2 - \|w(k)\|^2) - w^T(k)Cw(k)\right)\right] z_i(k), \qquad (2.13)
$$

for $k \geq 0$, where $i = 1, \ldots, n$.

Definition 2.2 *A point $w^* \in R^n$ is called an equilibrium of (2.12), if and only if*

$$
w^* = w^* + \eta\left[2Cw^* - Cw^*(w^*)^Tw^* - \left((w^*)^TCw^*\right)w^*\right].
$$

Clearly, the set of all equilibrium points of (2.12) is

$$
\{\text{all the eigenvectors with unit length}\} \cup \{0\}.
$$

Definition 2.3 *A bound M is called an ultimate bound of (2.12), if for any $w(0) \in S$, there exists a constant N so that $\|w(k)\|^2 < M$ for all $k > N$.*

For convenience in analysis, a lemma is presented as follows.

Lemma 2.6 *It holds that*

$$
\left[1 + \eta\left(\sigma(2 - s)\right)\right]^2 s \leq \frac{4}{27\eta\sigma} \cdot (1 + 2\eta\sigma)^3
$$

for all $s \in [0, 2 + 1/\eta\sigma]$.

It is easy to prove this lemma by Lemma 2.1

2.4.2 Invariant Set and Ultimate Bound

The system (2.12) is not globally convergent. Consider the same example as in Section 3, which is the one-dimensional case,

$$
w(k+1) = w(k) + \eta[2 - 2w^2(k)]w(k), \text{ for } k \geq 0.
$$

If $w(k) \geq 1 + 1/(2\eta)$, it holds that

$$2\eta w^2(k) - w(k) - 2\eta - 1 \geq 0.$$

Then

$$|w(k+1)| = w^2(k) \cdot \frac{2\eta w^2(k) - 2\eta - 1}{w(k)} \geq w^2(k).$$

$w(k)$ will clearly tend to infinity if $w(k) \geq 1 + 1/(2\eta)$. So, an important problem to address is whether an invariant set can be obtained to guarantee the nondivergence. This is a crucial problem to address to ensure the success of applications.

Lemma 2.7 *Suppose that* $\eta\sigma \leq 0.25$ *and* $0 < \|w(k)\|^2 \leq 1 + \dfrac{1}{2\sigma\eta}$ *, then*

$$1 + \eta(\lambda_i(2 - \|w(k)\|^2) - w^T(k)Cw(k)) > 0, \quad (i = 1, \ldots, n)$$

for all $k \geq 0$.

Proof : Given any $i(1 \leq i \geq n)$, if $2 \leq \|w(k)\|^2 < 1 + 1/2\eta\sigma$, then

$$1 + \eta\left(\lambda_i(2 - \|w(k)\|^2) - w^T(k)Cw(k)\right) \geq 1 + \eta\left(\sigma(2 - \|w(k)\|^2) - \sigma\|w^T\|^2\right) > 0,$$

for all $k \geq 0$. If $\|w(k)\|^2 < 2$, then

$$1 + \eta\left(\lambda_i(2 - \|w(k)\|^2) - w^T(k)Cw(k)\right) \geq 1 - \eta w^T(k)Cw(k) > 1 - 2\eta\sigma > 0,$$

for all $k \geq 0$. The proof is completed.

Theorem 2.4 *Denote*

$$S = \left\{ w(k) | w(k) \in R^n, \|w(k)\|^2 \leq 1 + \frac{1}{2\sigma\eta} \right\}.$$

If

$$\eta\sigma \leq 0.25,$$

then S *is an invariant set of (2.12).*

Proof: From Lemma 2.7, if $2 \leq \|w(k)\|^2 \leq 1 + \dfrac{1}{2\sigma\eta}$, then

$$0 < 1 + \eta\left(\lambda_i(2 - \|w(k)\|^2) - w^T(k)Cw(k)\right) < 1, \quad (i = 1, \ldots, n)$$

for $k \geq 0$. From (1.7), (2.13), it follows that

$$
\begin{aligned}
\|w(k+1)\|^2 &= \sum_{i=1}^{n}\left[1 + \eta\left(\lambda_i(2 - \|w(k)\|^2) - w^T(k)Cw(k)\right)\right]^2 z_i^2(k) \\
&\leq \left[1 + \eta\left(\lambda_p(2 - \|w(k)\|^2) - w^T(k)Cw(k)\right)\right]^2 \|w(k)\|^2 \\
&\leq 1 + \frac{1}{2\sigma\eta}.
\end{aligned}
$$

If $\|w(k)\|^2 < 2$, then

$$
\begin{aligned}
\|w(k+1)\|^2 &= \sum_{i=1}^{n} \left[1 + \eta \left(\lambda_i (2 - \|w(k)\|^2) - w^T(k)Cw(k) \right) \right]^2 z_i^2(k) \\
&\leq \left[1 + \eta \left(\sigma(2 - \|w(k)\|^2) - w^T(k)Cw(k) \right) \right]^2 \|w(k)\|^2 \\
&\leq \left[1 + \eta \left(\sigma(2 - \|w(k)\|^2) \right) \right]^2 \|w(k)\|^2 \\
&\leq \max_{0<s<2} \left\{ [1 + \eta(\sigma(2 - s))]^2 s \right\}.
\end{aligned}
$$

By Lemma (2.6),

$$
\|w(k+1)\|^2 < \frac{4}{27\eta\sigma} \cdot (1 + 2\eta\sigma)^3.
$$

If $\eta\sigma < 0.25$, then

$$
\frac{4(1 + 2\eta\sigma)^3}{27\sigma\eta} < \frac{1}{2\eta\sigma} < 1 + \frac{1}{2\eta\sigma}.
$$

The proof is completed.

Theorem 2.5 *Suppose that $\eta\sigma \leq 0.25$. The system (2.12) has an ultimate bound $\dfrac{2\sigma}{\sigma + \lambda_p}$; that is, if $w(0) \in S$ and $w(0) \notin V_\sigma^\perp$, then there exists a constant N, so that*

$$
\|w(k+1)\|^2 \leq \frac{2\sigma}{\sigma + \lambda_p} < 2,
$$

for all $k > N$.

Proof: From Lemma 2.7 and Theorem 2.4, if $w(0) \in S$, we have

$$
1 + \eta \left(\lambda_i(2 - \|w(k)\|^2) - w^T(k)Cw(k) \right) > 0, \quad (k \geq 0).
$$

If $2 \leq \|w(k)\|^2 < 1 + \dfrac{1}{2\eta\sigma}$, then

$$
\begin{aligned}
1 &+ \eta \left(\lambda_i(2 - \|w(k)\|^2) - w^T(k)Cw(k) \right) \\
&< 1 + \eta \left(\lambda_p(2 - \|w(k)\|^2) - w^T(k)Cw(k) \right) \\
&< 1,
\end{aligned}
$$

for all $k \geq 0$. Thus, it holds that

$$
\begin{aligned}
\|w(k+1)\|^2 &= \sum_{i=1}^{n} \left[1 + \eta \left(\lambda_i(2 - \|w(k)\|^2) - w^T(k)Cw(k) \right) \right]^2 z_i^2(k) \\
&\leq \left[1 + \eta \left(\lambda_p(2 - \|w(k)\|^2) - w^T(k)Cw(k) \right) \right]^2 \|w(k)\|^2 \\
&< \|w(k)\|^2,
\end{aligned}
$$

for all $k \geq 0$. If $\dfrac{2\sigma}{\sigma + \lambda_p} < \|w(k)\|^2 < 2$, then

$$1 + \eta \left(\lambda_i(2 - \|w(k)\|^2) - w^T(k)Cw(k)\right)$$
$$\leq 1 + \eta \left(2\sigma - (\sigma + \lambda_p)\|w(k)\|^2\right)$$
$$< 1,$$

for all $k \geq 0$. So, it follows

$$\|w(k+1)\|^2 = \sum_{i=1}^{n} \left[1 + \eta \left(\lambda_i(2 - \|w(k)\|^2) - w^T(k)Cw(k)\right)\right]^2 z_i^2(k)$$
$$\leq \left[1 + \eta \left(2\sigma - (\sigma + \lambda_p)\|w(k)\|^2\right)\right]^2 \|w(k)\|^2$$
$$< \|w(k)\|^2,$$

for all $k \geq 0$. If $0 < \|w(k)\|^2 \leq \dfrac{2\sigma}{\sigma + \lambda_p}$, then

$$\|w(k+1)\|^2 = \sum_{i=1}^{n} \left[1 + \eta \left(\lambda_i(2 - \|w(k)\|^2) - w^T(k)Cw(k)\right)\right]^2 z_i^2(k)$$
$$\leq \left[1 + \eta \left(2\sigma - (\sigma + \lambda_p)\|w(k)\|^2\right)\right]^2 \|w(k)\|^2$$
$$\leq \left[1 + \frac{0.25}{\sigma} \left(2\sigma - (\sigma + \lambda_p)\|w(k)\|^2\right)\right]^2 \|w(k)\|^2$$
$$\leq \left[1.5 - 0.25\frac{\sigma + \lambda_p}{\sigma}\|w(k)\|^2\right]^2 \|w(k)\|^2$$
$$\leq \frac{2\sigma}{\sigma + \lambda_p}.$$

Therefore, if $w(0) \in S$ and $w(0) \notin V_\sigma^\perp$, then there must exist a constant N so that

$$\|w(k)\|^2 \leq \frac{2\sigma}{\sigma + \lambda_p} < 2,$$

for all $k > N$. The proof is completed.

The theorems above show the nondivergence of (2.12) is guaranteed if the learning rate satisfies a simple condition and the evolution of (2.12) is ultimately bounded.

2.4.3 Convergence Analysis

In this section, we will prove the trajectories, arising from points in the invariant set S, will converge to a unit eigenvector associated with the largest eigenvalue of the covariance matrix. To complete the proof, the following lemmas are first given.

Lemma 2.8 *Suppose that $\eta\sigma \leq 0.25$. If $w(0) \in S$ and $w(0) \neq 0$, then*

$$\eta w^T(k+1)Cw(k+1) \geq \gamma > 0,$$

for all $k \geq 0$, where

$$\gamma = \min\left\{(2\eta\lambda_p)^2(\eta\lambda_p)\frac{\lambda_p}{\sigma}, \eta w^T(0)Cw(0)\right\} < 1.$$

Proof: Since $w(0) \in S$, by Lemma (2.7), we have

$$1 + \eta\left(\lambda_i(2 - \|w(k)\|^2) - w^T(k)Cw(k)\right) > 0,$$

for all $k \geq 0$. From Theorem 2.5, if $2 \leq \|w(k)\|^2 \leq 1 + \dfrac{1}{2\eta\sigma}$, we have

$$\|w(k+1)\|^2 < \|w(k)\|^2, \quad (k \geq 0).$$

Case 1: $\dfrac{\lambda_p}{\sigma} < 2.$

If $\dfrac{\lambda_p}{\sigma} \leq \|w(k)\|^2 < 2$, from (1.8) and (2.13), it follows that

$$
\begin{aligned}
&\eta w^T(k+1)Cw(k+1) \\
=\;& \eta \sum_{i=1}^{n} \lambda_i z_i^2(k+1) \\
=\;& \eta \sum_{i=1}^{p} \left[1 + \eta\left(\lambda_i(2 - \|w(k)\|^2) - w^T(k)Cw(k)\right)\right]^2 \lambda_i z_i^2(k) \\
\geq\;& \eta \sum_{i=1}^{p} \left[1 + \eta\left(2\lambda_p - \sigma\|w(k)\|^2 - \sigma\|w(k)\|^2\right)\right]^2 \lambda_i z_i^2(k) \\
\geq\;& \eta \left[2\eta\lambda_p + \left(1 - 2\eta\sigma\|w(k)\|^2\right)\right]^2 w^T(k)Cw(k) \\
\geq\;& (2\eta\lambda_p)^2 \cdot \eta w^T(k)Cw(k) \\
\geq\;& (2\eta\lambda_p)^2(\eta\lambda_p)\frac{\lambda_p}{\sigma}
\end{aligned}
$$
(2.14)

for $k \geq 0$.

If $\|w(k)\|^2 \leq \dfrac{\lambda_p}{\sigma}$, then

$$
\begin{aligned}
&\eta w^T(k+1)Cw(k+1) \\
=\;& \eta \sum_{i=1}^{p} \left[1 + \eta\left(\lambda_i - w^T(k)Cw(k)\right)\right]^2 \lambda_i z_i^2(k) \\
\geq\;& \eta \left[1 + \eta\left(2\lambda_p - \sigma\|w(k)\|^2 - \sigma\|w(k)\|^2\right)\right]^2 w^T(k)Cw(k) \\
\geq\;& \eta w^T(k)Cw(k)
\end{aligned}
$$
(2.15)

for $k \geq 0$.

Case 2: $\dfrac{\lambda_p}{\sigma} > 2$.

Clearly, if $2 \leq \|w(k)\|^2 < \dfrac{\lambda_p}{\sigma}$, it holds that

$$\|w(k+1)\|^2 < \|w(k)\|^2, \quad (k \geq 0).$$

If $0 < \|w(k)\|^2 < 2$, from (2.15), it follows that

$$\eta w^T(k+1)Cw(k+1) \geq \eta w^T(k)Cw(k), \quad (k \geq 0). \tag{2.16}$$

Let

$$\gamma = \min \left\{ 4(\eta \lambda_p)^3 \frac{\lambda_p}{\sigma}, \eta w^T(0)Cw(0) \right\},$$

where $0 < \gamma < 1$. It follows from (2.14), (2.15), and (2.16) that

$$\eta w^T(k+1)Cw(k+1) \geq \gamma > 0,$$

for all $k \geq 0$. This completes the proof.

Lemma 2.9 *If $w(0) \in S$ and $w(0) \notin V_\sigma^\perp$, then there exist constants $\theta_1 > 0$, $\Pi_1 \geq 0$, and $d > 0$ such that*

$$\sum_{j=l+1}^{n} z_j^2(k) \leq \Pi_1 \cdot e^{-\theta_1 k},$$

for all $k > N$, where

$$\theta_1 = ln \left(\frac{\sigma + \lambda_p + 2\eta\sigma\lambda_p}{\sigma + \lambda_p + 2\eta\lambda_{m+1}\lambda_p} \right)^2 > 0.$$

Proof: Since $w(0) \notin V_\sigma^\perp$, there must exist some $i(1 \leq i \leq m)$ such that $z_i(0) \neq 0$. Without loss of generality, assume that $z_1(0) \neq 0$.

By Theorem 2.4, it follows that $w(k) \in S$ for all $k \geq 0$. From Lemma 2.7, it holds that

$$1 + \eta \left(\lambda_i(2 - \|w(k)\|^2) - w^T(k)Cw(k) \right) > 0,$$

for $k \geq 0$. From (2.13), for each $j(m+1 \leq j \leq n)$, it follows that

$$\left[\frac{z_j(k+1)}{z_1(k+1)} \right]^2 = \left[\frac{1 + \eta[\lambda_i(2 - \|w(k)\|^2) - w^T(k)Cw(k)]}{1 + \eta[\sigma(2 - \|w(k)\|^2) - w^T(k)Cw(k)]} \right]^2 \cdot \left[\frac{z_j(k)}{z_1(k)} \right]^2.$$

From Theorem 2.5, then

$$\|w(k+1)\|^2 \leq \frac{2\sigma}{\sigma + \lambda_p} < 2,$$

for all $k > N$. So, it follows that

$$
\begin{aligned}
\left[\frac{z_j(k+1)}{z_1(k+1)}\right]^2 &\leq \left[\frac{1+\eta[\lambda_i(2-\|w(k)\|^2)]}{1+\eta[\sigma(2-\|w(k)\|^2)]}\right]^2 \cdot \left[\frac{z_j(k)}{z_1(k)}\right]^2 \\
&\leq \left[\frac{1+\eta\left[\lambda_i\left(2-\dfrac{2\sigma}{\sigma+\lambda_p}\right)\right]}{1+\eta\left[\sigma\left(2-\dfrac{2\sigma}{\sigma+\lambda_p}\right)\right]}\right]^2 \cdot \left[\frac{z_j(k)}{z_1(k)}\right]^2 \\
&= \left[\frac{\sigma+\lambda_p+2\eta\lambda_{m+1}\lambda_p}{\sigma+\lambda_p+2\eta\sigma\lambda_p}\right]^2 \cdot \left[\frac{z_j(k)}{z_1(k)}\right]^2 \\
&= \left(\frac{\sigma+\lambda_p+2\eta\lambda_{m+1}\lambda_p}{\sigma+\lambda_p+2\eta\sigma\lambda_p}\right)^{2(k+1)} \cdot \left[\frac{z_j(0)}{z_1(0)}\right]^2 \\
&= \left[\frac{z_j(0)}{z_1(0)}\right]^2 \cdot e^{-\theta_1(k+1)},
\end{aligned}
$$

for all $k > N$.

Since $w(k) \in S$, $z_1(k)$ must be bounded; that is, there exists a constant $d > 0$ such that $z_1^2(k) \leq d$. Then,

$$
\sum_{j=m+1}^{n} z_j^2(k) = \sum_{j=m+1}^{n} \left[\frac{z_j(k)}{z_1(k)}\right]^2 \cdot z_1^2(k) \leq \Pi_1 e^{-\theta_1 k},
$$

for all $k > N$, where

$$
\Pi_1 = d \sum_{j=m+1}^{n} \left[\frac{z_j(0)}{z_1(0)}\right]^2 \geq 0.
$$

This completes the proof.

Lemma 2.10 *If $w(0) \in S$ and $w(0) \notin V_\sigma^\perp$, then there exists constants $\theta_2 > 0$ and $\Pi_2 > 0$ such that*

$$
|\sigma(2-\|w(k)\|^2) - w^T(k+1)Cw(k+1)| \leq k \cdot \Pi_2 \cdot \left[e^{-\theta_2(k+1)} + \max\left\{e^{-\theta_2 k}, e^{-\theta_1 k}\right\}\right],
$$

for all $k > N$, where

$$
\begin{cases}
\theta_2 = -\ln\delta, \quad (0 < \delta < 1), \\
\delta = \max\left\{(1-\gamma)^2, 2\eta\sigma\right\}, \\
\gamma = \min\left\{(2\eta\lambda_p)^2(\eta\lambda_p)(\frac{\lambda_p}{\sigma}), \eta w^T(0)Cw(0)\right\},
\end{cases}
$$

and $0 < \delta < 1$, $0 < \gamma < 1$.

Proof: From (1.8), (2.13), we have

$$w^T(k+1)Cw(k+1)$$
$$= \sum_{i=1}^{n} \lambda_i \left[1 + \eta \left(\lambda_i(2 - \|w(k)\|^2) - w^T(k)Cw(k)\right)\right]^2 z_i^2(k)$$
$$= \sum_{i=1}^{n} \lambda_i \left[1 + \eta \left(\sigma(2 - \|w(k)\|^2) - w^T(k)Cw(k)\right)\right]^2 z_i^2(k)$$
$$+ \sum_{i=m+1}^{n} \lambda_i \left[1 + \eta \left(\lambda_i(2 - \|w(k)\|^2) - w^T(k)Cw(k)\right)\right]^2 z_i^2(k)$$
$$- \sum_{i=m+1}^{n} \lambda_i \left[1 + \eta \left(\sigma(2 - \|w(k)\|^2) - w^T(k)Cw(k)\right)\right]^2 z_i^2(k)$$
$$= \left[1 + \eta \left(\sigma(2 - \|w(k)\|^2) - w^T(k)Cw(k)\right)\right]^2 \cdot w^T(k)Cw(k) - Q(k)$$

for $k \geq 0$, where

$$Q(k) = \eta \sum_{i=m+1}^{n} \lambda_i \left(\sigma - \lambda_i\right)\left(2 - \|w(k)\|^2\right)$$
$$\times \left[2 + \eta \left((\sigma + \lambda_i)(2 - \|w(k)\|^2) - 2w^T(k)Cw(k)\right)\right] z_i^2(k).$$

Then

$$\sigma(2 - \|w(k)\|^2) - w^T(k+1)Cw(k+1)$$
$$= \left[\sigma(2 - \|w(k)\|^2) - w^T(k)Cw(k)\right]$$
$$\times \left[\left(1 - \eta w^T(k)Cw(k)\right)^2 - \eta^2 \sigma(2 - \|w(k)\|^2)w^T(k)Cw(k)\right]$$
$$+ Q(k),$$

for $k \geq 0$. Denote

$$V(k) = \left|\sigma(2 - \|w(k)\|^2) - w^T(k)Cw(k)\right|,$$

for $k \geq 0$. By the invariance of S, $w(k) \in S$ for $k \geq 0$. It follows that,

$$V(k+1)$$
$$\leq V(k) \cdot \left|\left(1 - \eta w^T(k)Cw(k)\right)^2 - \eta^2 \sigma(2 - \|w(k)\|^2)w^T(k)Cw(k)\right|$$
$$+ |Q(k)|$$
$$\leq \max\left\{\left(1 - \eta w^T(k)Cw(k)\right)^2, \eta^2 \sigma(2 - \|w(k)\|^2)w^T(k)Cw(k)\right\} \cdot V(k)$$
$$+ |Q(k)|$$
$$\leq \max\left\{\left(1 - \eta w^T(k)Cw(k)\right)^2, \eta \sigma(2 - \|w(k)\|^2)\right\} \cdot V(k) + |Q(k)|.$$

From Theorem 2.5, it follows that

$$V(k+1) \le \max\left\{\left(1 - \eta w^T(k)Cw(k)\right)^2, 2\eta\sigma\right\} \cdot V(k) + |Q(k)|,$$

for $k > N$. Denote

$$\gamma = \min\left\{(2\eta\lambda_p)^2(\eta\lambda_p)\frac{\lambda_p}{\sigma}, \eta w^T(0)Cw(0)\right\}.$$

Clearly, $0 < \gamma < 1$. By Lemma 2.9, it holds that

$$\gamma \le \eta w^T(k)Cw(k) \le 1.$$

Denote

$$\delta = \max\left\{(1-\gamma)^2, 2\eta\sigma\right\}.$$

Clearly, $0 < \delta < 1$. Then

$$V(k+1) \le \delta \cdot V(k) + |Q(k)|, k \ge N.$$

By Lemma 2.8,

$$
\begin{aligned}
|Q(k)| &\le \eta \sum_{j=m+1}^{n} 2\lambda_j (\sigma - \lambda_j) \left[2 + 2\eta(\sigma + \lambda_j)\right] z_j^2(k) \\
&\le \eta \sum_{j=m+1}^{n} 4\left[\eta\lambda_j(\sigma^2 - \lambda_j^2) + \lambda_j(\sigma - \lambda_j)\right] z_j^2(k) \\
&\le \eta \sum_{j=m+1}^{n} 4\left(\eta\lambda_j\sigma^2 + 2\lambda_j\sigma\right) z_j^2(k) \\
&\le 4\sigma(\eta\sigma)(\eta\sigma + 2) \sum_{j=m+1}^{n} z_j^2(k) \\
&\le 12\sigma \cdot \sum_{j=m+1}^{n} z_j^2(k) \\
&\le 12\sigma\Pi_1 \cdot e^{-\theta_1 k}
\end{aligned}
$$

for all $k \ge N$. Then

$$
\begin{aligned}
V(k+1) &\le \delta^{k+1}V(0) + 12\sigma\Pi_1 \cdot \sum_{r=0}^{k}\left(\delta e^{\theta_1}\right)^r e^{-\theta_1 k} \\
&\le \delta^{k+1}V(0) + 12k\sigma\Pi_1 \cdot \max\left\{\delta^k, e^{-\theta_1 k}\right\} \\
&\le k \cdot \Pi_2 \cdot \left[e^{-\theta_2(k+1)} + \max\left\{e^{-\theta_2 k}, e^{-\theta_1 k}\right\}\right],
\end{aligned}
$$

where $\theta_2 = -\ln\delta > 0$ and

$$\Pi_2 = \max\left\{\left|2\sigma - w^T(0)Cw(0)\right|, 12\sigma\Pi_1\right\} > 0.$$

The proof is completed.

Lemma 2.11 *Suppose there exists constants $\theta > 0$ and $\Pi > 0$ such that*

$$\eta \left| \left(\sigma(2 - \|w(k)\|^2) - w^T(k)Cw(k) \right) z_i(k) \right| \leq k \cdot \Pi e^{-\theta k}, (i = 1, \ldots, m)$$

for $k \geq N$. Then

$$\lim_{k \to +\infty} z_i(k) = z_i^*, \quad (i = 1, \ldots, m),$$

where $z_i^*(i = 1, \ldots, m)$ are constants.

Proof: Given any $\epsilon > 0$, there exists a $K \geq 1$ such that

$$\frac{\Pi K e^{-\theta K}}{(1 - e^{-\theta})^2} \leq \epsilon.$$

For any $k_1 > k_2 \geq K$, it follows that

$$
\begin{aligned}
|z_i(k_1) - z_i(k_2)| &= \left| \sum_{r=k_2}^{k_1-1} [z_i(r+1) - z_i(r)] \right| \\
&\leq \eta \sum_{r=k_2}^{k_1-1} \left| \left(\sigma(2 - \|w(k)\|^2) - w^T(r)Cw(r) \right) z_i(r) \right| \\
&\leq \Pi \sum_{r=k_2}^{k_1-1} r e^{-\theta r} \\
&\leq \Pi \sum_{r=K}^{+\infty} r e^{-\theta r} \\
&\leq \Pi K e^{-\theta K} \cdot \sum_{r=0}^{+\infty} r \left(e^{-\theta} \right)^{r-1} \\
&= \frac{\Pi K e^{-\theta K}}{(1 - e^{-\theta})^2} \\
&\leq \epsilon, \quad (i = 1, \ldots, m).
\end{aligned}
$$

This shows that each sequence $\{z_i(k)\}$ is a *Cauchy sequence*. By *Cauchy Convergence Principle*, there must exist constants $z_i^*(i = 1, \ldots, m)$ such that

$$\lim_{k \to +\infty} z_i(k) = z_i^*, \quad (i = 1, \ldots, m).$$

This completes the proof.

Theorem 2.6 *Suppose that*

$$\eta \sigma \leq 0.25.$$

If $w(0) \in S$ and $w(0) \notin V_\sigma^\perp$, then the trajectory of (2.12) starting from $w(0)$ will converge to a unit eigenvector associated with the largest eigenvalue of the covariance matrix C.

Proof: By Lemma 2.9, there exists constants $\theta_1 > 0$ and $\Pi_1 \geq 0$ such that

$$\sum_{j=m+1}^{n} z_j^2(k) \leq \Pi_1 e^{-\theta_1 k},$$

for all $k \geq N$. By Lemma 2.10, there exists constants $\theta_2 > 0$ and $\Pi_2 > 0$ such that

$$\left| \sigma(2 - \|w(k)\|^2) - w^T(k+1)Cw(k+1) \right| \leq k \cdot \Pi_2 \cdot \left[e^{-\theta_2(k+1)} + \max\left\{ e^{-\theta_2 k}, e^{-\theta_1 k} \right\} \right]$$

for all $k \geq N$.

Obviously, there exists constants $\theta > 0$ and $\Pi > 0$ such that

$$\eta \left| \left(\sigma(2 - \|w(k)\|^2) - w^T(k)Cw(k) \right) z_i(k) \right| \leq k \cdot \Pi e^{-\theta k}, \quad (i = 1, \ldots, m)$$

for $k \geq N$.

Using Lemma 2.9 and Lemma 2.11, it follows that

$$\begin{cases} \lim_{t \to +\infty} z_i(k) = z_i^*, & (i = 1, \ldots, m) \\ \lim_{t \to +\infty} z_i(k) = 0, & (i = m+1, \ldots, n). \end{cases}$$

Thus,

$$\lim_{t \to +\infty} w(k) = \sum_{i=1}^{m} z_i^* v_i \in V_\sigma. \tag{2.17}$$

From (2.12), after the system becomes stable, it follows that

$$\lim_{k \to \infty} 2Cw(k) = \lim_{k \to \infty} \left[Cw(k)w^T(k)w(k) - \left(w^T(k)Cw(k) \right) w(k) \right]. \tag{2.18}$$

Substitute (2.17) into (2.18), we get

$$\sum_{i=1}^{m} \sigma z_i^* v_i = \sum_{i=1}^{m} \sigma z_i^* v_i \sum_{i=1}^{m} (z_i^*)^2 + \sum_{i=1}^{m} \sigma(z_i^*)^2 \sum_{i=1}^{m} z_i^* v_i.$$

It is easy to see that

$$\sum_{i=1}^{m} (z_i^*)^2 = 1.$$

The proof is completed.

The above theorem requires that the initial $w(0) \notin V_\sigma^\perp$ so that there is at least one $z_i(0) \neq 0 (1 \leq i \leq m)$. Since any small disturbance can result in $w(0) \notin V_\sigma^\perp$, the condition $w(0) \notin V_\sigma^\perp$ is easy to meet in practical applications. Therefore, as long as the learning rate satisfies a simple condition, almost all trajectories starting from S will converge to a unit eigenvector associated with the largest eigenvalue of the covariance matrix of C.

2.4.4   Simulations and Discussions

Simulations in this section show the convergence of Xu's algorithm with a constant learning rate. First, we randomly generate a 6×6 symmetric nonnegative definite matrix as

$$C = \begin{bmatrix} 0.1712 & 0.1538 & 0.097645 & 0.036741 & 0.07963 & 0.12897 \\ 0.1538 & 0.13855 & 0.087349 & 0.033022 & 0.072609 & 0.11643 \\ 0.097645 & 0.087349 & 0.067461 & 0.032506 & 0.043641 & 0.070849 \\ 0.036741 & 0.033022 & 0.032506 & 0.019761 & 0.016771 & 0.025661 \\ 0.07963 & 0.072609 & 0.043641 & 0.016771 & 0.041089 & 0.062108 \\ 0.12897 & 0.11643 & 0.070849 & 0.025661 & 0.062108 & 0.098575 \end{bmatrix}.$$

The maximum eigenvalue $\sigma = 0.5099$. The invariant set S can be identified as

$$S = \left\{ w(k) | w(k) \in R^2, \|w(k)\|^2 \leq \frac{1}{2 \times \eta \times 0.5099} \right\},$$

where $\eta \leq 0.25/0.5099 \approx 0.4903$.

The algorithm (2.12) is not globally convergent. If the initial vector is

$$w(0) = \begin{bmatrix} 3.3500, 4.3000, 1.5000, 2.4000, 3.4000, 1.7000 \end{bmatrix}^T$$

and $\eta = 0.05$, $\|w(k)\|$ goes to infinity rapidly. Figure 2.6 shows the result. Theorem 2.6 shows that trajectories starting from $S - \{0\}$ will converge to an

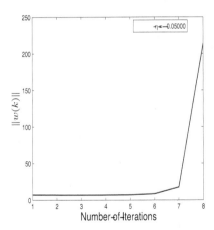

FIGURE 2.6
Divergence of (2.12).

eigenvector associated with the largest eigenvalue of the covariance matrix. The following simulations will confirm it. Selecting six initial vectors arbitrarily in $S - \{0\}$ is in Table 2.1.

TABLE 2.1 The $Norm^2$ of Initial Vector

2.5214	3.1887	1.3608	0.0044	5.3374	0.1287

Figure 2.7 (left) shows the six trajectories converging to the unit hyperspherical plane. The evolution result with the same initial vector and four different learning rates is presented in Figure 2.7 (right).

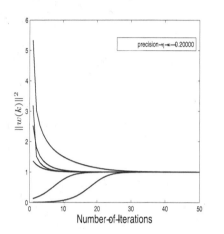 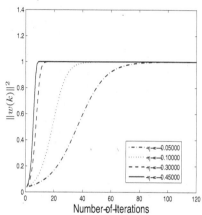

FIGURE 2.7
Convergence of (2.12) with different initial vectors (left) and with different learning rates (right).

To measure the convergent direction, we compute the norm of $w(k)$ and the direction cosine at kth update by [32], [31]:

$$\text{Direction Cosine}(k) = \frac{|w^T(k) \cdot \phi|}{\|w(k)\| \cdot \|\phi\|},$$

where ϕ is the true eigenvector associated with the largest eigenvalue of C. Then, we select arbitrarily an initial vector as

$$w(0) = \begin{bmatrix} 0.0900, 0.0400, 0.0950, 0.0900, 0.0840, 0.0600 \end{bmatrix}^T.$$

Figure 2.8 (left) shows the convergence of the direction cosine and the convergence of $\|w(k)\|$. The right one shows the components of w converge to

$$w^* = \begin{bmatrix} 0.58584, 0.56139, 0.39833, 0.14884, 0.24411, 0.31816 \end{bmatrix}^T.$$

This algorithm is more suitable for online learning with high performance. The online evolution behavior is illustrated by extracting a principal feature

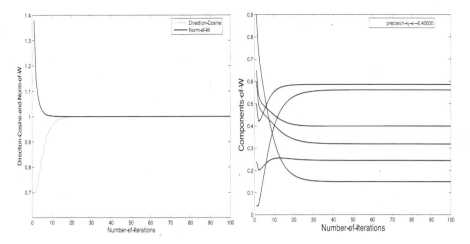

FIGURE 2.8
Convergence of direction cosine (left) and components of w (right).

from Lenna picture in Figure 1.5. By numerical discretization procedure, the Xu's algorithm (2.12) with a constant learning rate could be rewritten as follows that

$$w(k+1) = w(k) + \eta(2C_k w(k) - w(k)^T w(k) C_k w(k) - (w(k)^T C_k w(k))w(k)),$$

for $k \geq 0$, where C_x is a online observation sequence. The online observation sequence $\{C_k\}$ (1.10) will be used to train the system with $\lim_{k \to \infty} C_k = C, (k > 0)$. The reconstructed image is presented in Figure 2.9 with SNR 32.175.

2.4.5 Conclusion

The convergence of Xu's LMSER algorithm, in the one unit case (2.11), is interpreted by using the DDT method. An invariant set and an ultimate bound are derived to guarantee the nondivergence of Xu's DDT system. It is rigorously proven, if the learning rate satisfy some simple condition, all trajectories, starting from points in this invariant set, converge to the eigensubspace, which is spanned by the eigenvectors associated with the largest eigenvalue of the covariance matrix.

2.5 Comparison of Oja's Algorithm and Xu's Algorithm

In this section, by comparing Oja's algorithm with Xu's algorithm, the computational complexity and convergence rate of both algorithms are discussed.

FIGURE 2.9
The reconstructed image of the Lenna image.

It could be observed, on the whole, the Xu's algorithm converges faster or diverges more rapidly with time, albeit at a cost of higher computational complexity.

In Section 3, the Oja's DDT algorithm with a constant learning rate is analyzed in detail. Some important results have been presented. The Oja's DDT algorithm can be presented as follows:

$$w(k+1) = w(k) + \eta[Cw(k) - (w^T(k)Cw(k))w(k)],$$

for $k \geq 0$. Clearly, Xu's algorithm (2.12) approximates Oja's one unit algorithm when the normal of the weight vector approaches 1. However, some different dynamical behaviors can be shown.

On the one hand, Xu's algorithm has a different attractive domain. Given a same input sequence, Oja's algorithm requires $\eta\sigma < 0.8899$ and its attractive domain is

$$S_{oja} = \left\{ w(k) | w(k) \in R^n, w^T(k)Cw(k) < \frac{1}{\eta} \right\}.$$

Xu's algorithm requires $\eta\sigma < 0.25$ with the attractive domain

$$S_{xu} = \left\{ w(k) | w(k) \in R^n, \|w(k)\|^2 < 1 + \frac{1}{2\eta\sigma} \right\}.$$

On the other hand, we find Xu's algorithm converges faster on the whole, though Xu's algorithm (2.12) has a larger computation in each iteration. In Section 3, the coefficients $z_i(k)$ of weight about Oja's algorithm is presented as

$$z_i(k+1)_{oja} = \left[1 + \eta \left(\lambda_i - w^T(k)Cw(k) \right) \right] z_i(k),$$

for $k \geq 0$, where $i = 1, \ldots, n$. Suppose $w(k)$ is same, with the same learning rate. From (2.13), we have

$$
\left| \frac{z_i(k+1)_{xu} - z_i(k+1)}{z_i(k+1)_{oja} - z_i(k+1)} \right| = \left| \frac{\left[\eta \left(\lambda_i (2 - \|w(k)\|^2) - w^T(k)Cw(k) \right) \right] z_i(k)}{\left[\eta \left(\lambda_i - w^T(k)Cw(k) \right) \right] z_i(k)} \right|
$$

$$
= \left| 1 + \frac{\lambda_i(1 - \|w(k)\|^2)}{\lambda_i - w^T(k)Cw(k)} \right|.
$$

Clearly, it holds that

$$
\left| \frac{z_i(k+1)_{xu} - z_i(k+1)}{z_i(k+1)_{oja} - z_i(k+1)} \right| = \left| 1 + \frac{\lambda_i(1 - \|w(k)\|^2)}{\lambda_i - w^T(k)Cw(k)} \right| > 1,
$$

for $w^T(k)Cw(k) < \lambda_p$ or $w^T(k)Cw(k) > \sigma$. So, in this range, Xu's algorithm converges faster. However, if $\lambda_p < w^T(k)Cw(k) < \sigma$, it can be observed that the evolution rate of both algorithms interlace and trend to the same rate. Figure 2.10 (right) shows the result. Furthermore, it is not difficult to get

$$
\begin{cases} \sigma > w^T(k)Cw(k), & \text{for } \|w(k)\|^2 < 1, \\ \lambda_p < w^T(k)Cw(k), & \text{for } \|w(k)\|^2 > 1. \end{cases}
$$

Thus, it holds that

$$
\begin{cases} \left| \dfrac{z_1(k+1)_{xu} - z_1(k+1)}{z_1(k+1)_{oja} - z_1(k+1)} \right| = \left| 1 + \dfrac{\sigma(1 - \|w(k)\|^2)}{\sigma - w^T(k)Cw(k)} \right| > 1, & \text{for } \|w(k)\|^2 < 1, \\[4mm] \left| \dfrac{z_p(k+1)_{xu} - z_p(k+1)}{z_p(k+1)_{oja} - z_p(k+1)} \right| = \left| 1 + \dfrac{\lambda_p(1 - \|w(k)\|^2)}{\lambda_p - w^T(k)Cw(k)} \right| > 1, & \text{for } \|w(k)\|^2 > 1. \end{cases}
$$

Therefore, in at least one direction, Xu's algorithm converges faster if $\lambda_p < w^T(k)Cw(k) < \sigma$ and by a number of simulations, it can be observed that, on the whole, Xu's algorithm converges faster. Consider a simple example, a input matrix is randomly generated as follows. Figure 2.10 shows the evolution result of both algorithms.

$$
C = \begin{bmatrix} 0.9267 & 0.0633 & -0.2633 \\ 0.0633 & 0.1900 & 0.1267 \\ -0.2633 & 0.1267 & 0.9000 \end{bmatrix}.
$$

2.6 Conclusion

In this chapter, Oja's algorithm and Xu's LMSER algorithm are studied by using DDT method. Some invariant sets are obtained so that the nondivergence of algorithms can be guaranteed. It is rigorously proved, if the learning

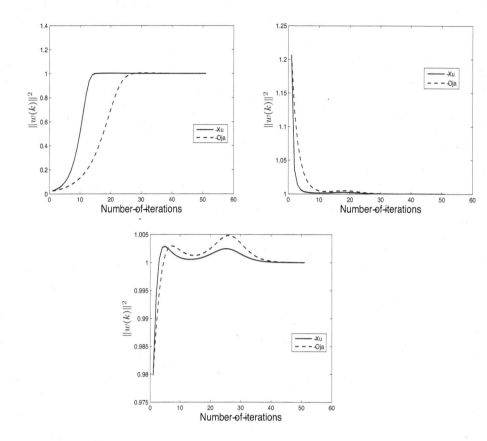

FIGURE 2.10
Comparison of evolution rate of Xu's and Oja's algorithm with a same learning
rate $\eta = 0.3$.

rate satisfy some simple condition, all trajectories, starting from points in this
invariant set, converge to the eigensubspace, which is spanned by the eigenvectors associated with the largest eigenvalue of the covariance matrix. The
most important result is that it confirms that the use of a constant learning
rate for some PCA learning algorithms can achieve convergence. Simulation
studies are carried out to further illustrate the theory.

3

PCA Learning Algorithms with Adaptive Learning Rates

3.1 Introduction

In PCA learning algorithms, learning rate is crucial. It not only influences the convergence speed of PCA algorithms but also determines whether this algorithm is convergent. Clearly, the DCT method requires that the learning rate approaches to zero. In Chapter 2, by using DDT method, it is shown that Oja's and Xu's LMSER's PCA algorithms with constant learning rates are locally convergent.

In this chapter, an non-zero-approaching adaptive learning rate is proposed for PCA learning algorithms. The convergence of Oja's learning algorithm with the adaptive learning rate is investigated in detail via the DDT method. First, the original SDT algorithm is transformed into the corresponding DDT algorithm. Then, the global boundedness of the DDT algorithm is analyzed. Under the conditions of nondivergence, global convergence is proved. The convergence of the DDT algorithm can shed some light on that of the original algorithm.

The rest of this chapter is organized as follows. In Section 2, a non-zero-approaching adaptive learning rate is proposed for PCA learning algorithm. Section 3 gives the problem formulation. And preliminaries for convergence proof are described. In Section 4, the convergence results are obtained. The lower and upper bounds of the weight evolution are found. The global convergence is proved in detail. Simulations are carried out in Section 5. Conclusions are drawn in Section 6.

3.2 Adaptive Learning Rate

With the linear neural network (1.1) and input sequence (1.2), the Oja's PCA learning algorithm (2.1) can be rewritten as

$$w(k + 1) = w(k) + \eta(k)[C_{x(k)}w(k) - w^T(k)C_{x(k)}w(k)w(k)], \qquad (3.1)$$

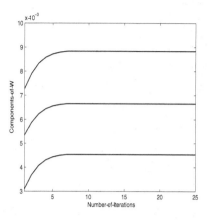

FIGURE 3.1

Converge to an incorrect direction with $\eta(k) = \dfrac{1}{2^k + 1}$.

where $\eta(k) > 0$ is the learning rate, and $C_{x(k)} = x(k)x^T(k)$ is an online observation of the covariance matrix defined by $C = E\left[x(k)x^T(k)\right]$.

Before we propose the non-zero-approaching adaptive learning rate, let us use a simple example to see how a zero-approaching learning rate affects the convergence of (3.1). Consider the following simple example,

$$w(k+1) = w(k)\left(1 + \eta(k)(1 - w(k)^2)\right), \qquad (3.2)$$

for $k \geq 0$. If we choose $\eta(k) = \dfrac{1}{k+1}$, then $\|\ w(k+1)\ \| \to \infty$ as $\|\ w(0)\ \| > 2$ [210]. If we choose $\eta(k) = \dfrac{1}{2^k + 1}$ and the initial weight vector $w(0) = [0.0053408, 0.0072711, 0.0030929]^T$, the algorithm (3.2) converges to an incorrect vector $w^* = [0.0066514, 0.0088333, 0.0045339]^T$ in six iterations with $\|w^*\| = 0.0120$, see Figure 3.1.

The example shows that the algorithm will converge to an incorrect direction if the learning rate approaches to zero too fast.

In this chapter, we propose an adaptive learning rate that approaches a constant. The proposed adaptive learning rate is given as follows:

$$\eta(k) = \frac{\tilde{\xi}}{w^T(k)C_{x(k)}w(k)}, \qquad k \geq 0, \qquad (3.3)$$

where $C_{x(k)}$ is an online observation of C, and $\tilde{\xi}$ is a constant with $0 < \tilde{\xi} < 1$.

This adaptive learning rate approaches a positive constant. The convergence of (3.1) with the proposed learning rate will be studied by using the DDT method. It will show that (3.1) with the proposed learning rate is globally convergent. By global convergence, the selection of the initial vector of (3.1) will be quite simple, in fact, any nonzero vector can be taken as the initial vector. Moreover, the learning procedure will be speeded up due to the learning rate converges to a nonzero constant. Experiment results will further confirm that the algorithm (3.1), with the proposed non-zero-approaching learning rate, offers a high level of performance with an online observation sequence.

3.3 Oja's PCA Algorithms with Adaptive Learning Rate

Oja's PCA algorithm with adaptive learning rate is rewritten as

$$w(k+1) = w(k) + \frac{\tilde{\xi}}{w(k)^T C_{x(k)} w(k)} (C_{x(k)} w(k) - w(k)^T C_{x(k)} w(k) w(k)), \quad (3.4)$$

for all $k \leq 0$ and $0 < \tilde{\xi} < 1$.

We will use DDT method to indirectly study the convergence of (3.4). The DDT system characterizes the averaging evolution of the original one [210]. Taking the conditional expectation $E\{w(k+1)/w(0), w(i), i < k\}$ to both sides of (3.4), it follows that

$$E\{w(k+1)\} = E\left\{ w(k) + \frac{\tilde{\xi}}{w(k)^T C_{x(k)} w(k)} Y(k) \right\},$$

where

$$Y(k) = (C_{x(k)} w(k) - w(k)^T C_{x(k)} w(k) w(k)).$$

Clearly, the probability $P(Y(k))$ and $P(\frac{\tilde{\xi}}{w(k)^T C_{x(k)} w(k)})$ only depend on the distribution of the variable X. It is easy to see that $P(Y(k))$ is not affected by the value of the learning rate (3.3); that is, the value of the learning rates (3.3) gives no information on the value of $Y(k)$. Thus, the learning rate (3.3) and $Y(k)$ are independent each other. Thus, it holds that

$$E\{w(k+1)\} = E\{w(k)\} + E\left\{ \frac{\tilde{\xi}}{w(k)^T C_{x(k)} w(k)} \right\} E\{Y(k)\}.$$

Since

$$E\left\{ \frac{\tilde{\xi}}{w(k)^T C_{x(k)} w(k)} \right\} = \alpha \cdot \frac{\tilde{\xi}}{w(k)^T C w(k)} = \frac{\xi}{w(k)^T C w(k)},$$

where $\alpha > 0$, the corresponding DDT algorithm of (3.4) is as follows:

$$w(k+1) = w(k) + \frac{\xi}{w(k)^T C w(k)} (C w(k) - w(k)^T C w(k) w(k)), \quad (3.5)$$

for $k \geq 0$, where $0 < \xi < 0.8$. We will prove that (3.5) is globally convergent. From (1.7), (1.8), (3.5), we have

$$z_i(k+1) = \left[1 + \xi \left(\frac{\lambda_i}{w^T(k)Cw(k)} - 1\right)\right] z_i(k), \tag{3.6}$$

for $k \geq 0$, where $0 < \xi < 0.8$.

Then, a lemma is presented here, which will be used for the subsequent convergence analysis.

Lemma 3.1 *It holds that*

$$
\begin{cases}
\left[1 + \xi \left(\frac{\sigma}{s} - 1\right)\right]^2 s \geq 4(1 - \xi)\xi\sigma, \ for \ s > 0, \\[4mm]
\left[1 + \xi \left(\frac{\sigma}{s} - 1\right)\right]^2 s \leq max\left\{\sigma, \left[1 + \xi \left(\frac{\sigma}{s_*} - 1\right)\right]^2 s_*\right\}, \ for \ s \in [s_*, \sigma],
\end{cases}
$$

where $0 < \xi < 0.8$ *and* $s_* > 0$ *is a constant.*

Proof : Define a differentiable function

$$f(s) = \left[1 + \frac{\xi}{s}(\sigma - s)\right]^2 s$$

for $s > 0$. It follows that

$$\dot{f}(s) = \left[1 + \frac{\xi}{s}(\sigma - s)\right] \cdot \left[1 - \frac{\xi}{s}(\sigma + s)\right]$$

for $s > 0$. Denote

$$s^* = \frac{\xi\sigma}{1 - \xi}.$$

Then,

$$
\dot{f}(s) \begin{cases}
< 0, \ if \ 0 < s \leq s^* \\
= 0, \ if \ s = s^* \\
> 0, \ if \ s^* \leq s.
\end{cases}
$$

So, it shows that s^* must be the minimum point of the function $f(s)$ on the interval $(0, +\infty)$. Then, it holds that

$$\left[1 + \frac{\xi}{s}(\sigma - s)\right]^2 s \geq 4(1 - \xi)\xi\sigma,$$

for $s > 0$. And, it holds that

$$\left[1 + \frac{\xi}{s}(\sigma - s)\right]^2 s \leq max\left\{\sigma, \left[1 + \xi(\frac{\sigma}{s_*} - 1)\right]^2 s_*\right\}$$

for all $s \in [s_*, \sigma]$. The proof is complete.

3.4 Convergence Analysis of Oja's Algorithms with Adaptive Learning Rate

In this section, details associated with a convergence analysis of the proposed learning algorithm will be provided systematically.

3.4.1 Boundedness

First, we will analyze the boundedness of the system (3.5). The lower and upper bounds of the system evolution will be given in following theorems.

Lemma 3.2 *Suppose that* $0 < \xi < 0.8$, *if* $0 < w^T(k)Cw(k) < \lambda_p$, *then* $w^T(k)Cw(k)$ *is increasing.*

Proof : If $w^T(k)Cw(k) \leq \lambda_p$, it can be checked that

$$\left[1 + \xi\left(\frac{\lambda_i}{w^T(k)Cw(k)} - 1\right)\right]^2 \geq \left[1 + \xi\left(\frac{\lambda_p}{w^T(k)Cw(k)} - 1\right)\right]^2 \geq 1,$$

for $k \geq 0$. Then, from (3.5), (1.7), and (3.6), we have

$$
\begin{aligned}
w^T(k+1)Cw(k+1) &= \sum_{i=1}^{p}\left[1 + \xi\left(\frac{\lambda_i}{w^T(k)Cw(k)} - 1\right)\right]^2 \lambda_i z_i^2(k) \\
&\geq \left[1 + \xi\left(\frac{\lambda_p}{w^T(k)Cw(k)} - 1\right)\right]^2 \cdot w^T(k)Cw(k) \\
&\geq w^T(k)Cw(k)
\end{aligned}
$$

for $k \geq 0$. The proof is complete.

Theorem 3.1 *If* $w(0) \notin V_\sigma^\perp$, *it holds that*

$$w^T(k+1)Cw(k+1) \geq 4(1-\xi)\xi\lambda_p > 0,$$

for all $k \geq 0$, *where* $0 < \xi < 0.8$.

Proof: It can be checked that

$$\left[1 + \xi\left(\frac{\lambda_i}{w^T(k)Cw(k)} - 1\right)\right]^2 \geq \left[1 + \xi\left(\frac{\lambda_p}{w^T(k)Cw(k)} - 1\right)\right]^2 > 0,$$

for $k \geq 0$. From (3.5), (1.8), and (3.6), it follows that

$$
\begin{aligned}
w^T(k+1)Cw(k+1) &= \sum_{i=1}^{n} \lambda_i z_i^2(k+1) \\
&= \sum_{i=1}^{p} \lambda_i z_i^2(k+1) \\
&= \sum_{i=1}^{p} \left[1 + \xi\left(\frac{\lambda_i}{w^T(k)Cw(k)} - 1\right)\right]^2 \lambda_i z_i^2(k) \\
&\geq \left[1 + \xi\left(\frac{\lambda_p}{w^T(k)Cw(k)} - 1\right)\right]^2 \cdot w^T(k)Cw(k) \\
&\geq \min_{s>0}\left\{\left[1 + \xi\left(\frac{\lambda_p}{s} - 1\right)\right]^2 \cdot s\right\}
\end{aligned}
$$

for $k \geq 0$. By Lemma 3.1, it holds that

$$
w^T(k+1)Cw(k+1) \geq 4(1-\xi)\xi\lambda_p,
$$

for $k \geq 0$. This completes the proof.

The above theorem shows that, if $w(0) \neq 0$, the trajectory starting from $w(0)$ is lower bounded, then it will never converge to zero. Clearly, the zero vector must be an unstable equilibrium of (3.5).

Theorem 3.2 *There exists a constant $H > 0$ such that $w^T(k)Cw(k) \leq H$ for all $k \geq 0$.*

Proof: If $w^T(0)Cw(0) = 0$, then $w^T(k)Cw(k) = 0$ for all $k \geq 0$. Clearly, $w^T(k)Cw(k) \leq H$.

Next, assume that $w^T(0)Cw(0) \neq 0$. Define

$$
\bar{H} = \max\left\{\sigma, \left[1 + \xi\left(\frac{\sigma}{\lambda_p} - 1\right)\right]^2 \lambda_p, \left[1 + \xi\left(\frac{\sigma}{w^T(0)Cw(0)} - 1\right)\right]^2 w^T(0)Cw(0)\right\}.
$$

Given any positive constant H such that $H \geq \bar{H}$, we will prove that for any $k \geq 0$

$$
0 \leq w^T(k)Cw(k) \leq H
$$

implies that

$$
0 \leq w^T(k+1)Cw(k+1) \leq H.
$$

Two steps will be used to complete the proof.

Case 1. $\sigma \leq w^T(k)Cw(k) \leq H$. Then,

$$
\left[1 + \xi\left(\frac{\lambda_i}{w^T(k)Cw(k)} - 1\right)\right]^2 \leq 1, (i = 1, \dots, p).
$$

It follows that

$$
\begin{aligned}
w^T(k+1)Cw(k+1) &= \sum_{i=1}^{n} \lambda_i z_i^2(k+1) \\
&= \sum_{i=1}^{p} \lambda_i z_i^2(k+1) \\
&= \sum_{i=1}^{p} \lambda_i \left[1 + \xi\left(\frac{\lambda_i}{w^T(k)Cw(k)} - 1\right)\right]^2 z_i^2(k) \\
&\leq \sum_{i=1}^{p} \lambda_i z_i^2(k) \\
&= w^T(k)Cw(k) \\
&\leq H.
\end{aligned}
$$

Case 2. $0 \leq w^T(k)Cw(k) \leq \sigma$. It can be checked that

$$
\left[1 + \xi\left(\frac{\lambda_i}{w^T(k)Cw(k)} - 1\right)\right]^2 \leq \left[1 + \xi\left(\frac{\sigma}{w^T(k)Cw(k)} - 1\right)\right]^2,
$$

for $i = 1, \ldots, n$. It follows that

$$
\begin{aligned}
w^T(k+1)Cw(k+1) &= \sum_{i=1}^{p} \lambda_i \left[1 + \xi\left(\frac{\lambda_i}{w^T(k)Cw(k)} - 1\right)\right]^2 z_i^2(k) \\
&\leq \sum_{i=1}^{p} \lambda_i \left[1 + \xi\left(\frac{\sigma}{w^T(k)Cw(k)} - 1\right)\right]^2 z_i^2(k) \\
&= \left[1 + \xi\left(\frac{\sigma}{w^T(k)Cw(k)} - 1\right)\right]^2 \cdot w^T(k)Cw(k) \\
&\leq \max_{0 \leq s \leq \sigma}\left\{\left[1 + \frac{\xi}{s}(\sigma - s)\right]^2 s\right\}.
\end{aligned}
$$

If

$$
\lambda_p > s^* = \frac{\xi\sigma}{1 - \xi},
$$

by Lemma 3.1 and Lemma 3.2, it follows that

$$
w^T(k+1)Cw(k+1) \begin{cases} \leq \sigma, & \text{if } s^* < s < \sigma, \\ \leq \left[1 + \xi\left(\frac{\sigma}{w^T(0)Cw(0)} - 1\right)\right]^2 w^T(0)Cw(0), & \text{if } 0 < s < s^*. \end{cases}
$$

If $\lambda_p < s^*$, by Lemma 3.1 and Lemma 3.2,

$$w^T(k+1)Cw(k+1) \begin{cases} \leq \sigma, \text{ if } \lambda_p < s < \sigma, \\ \leq \left[1 + \xi\left(\dfrac{\sigma}{\lambda_p} - 1\right)\right]^2 \lambda_p, \text{ if } \lambda_p < s < s^* \\ \leq \left[1 + \xi\left(\dfrac{\sigma}{w^T(0)Cw(0)} - 1\right)\right]^2 w^T(0)Cw(0), \text{ if } 0 < s < \lambda_p. \end{cases}$$

Then,

$$w^T(k+1)Cw(k+1) \leq H.$$

The proof is complete.

3.4.2 Global Convergence

Consider the same one-dimensional equation example of Section 2 with our non-zero-approaching learning rate. We have

$$w(k+1) = w(k)\left(1 + \frac{\xi}{w^2(k)}(1 - w(k)^2)\right), \tag{3.7}$$

for $k \geq 0$, where $0 < \xi < 0.8$. Obviously, a set $\{0, -1, 1\}$ is equilibrium points of the system (3.7). If $|w(k)| < 1$, then $|w(k+1)|$ increases, and if $|w(k)| > 1$, then $|w(k+1)|$ decreases. Clearly, in the whole interval, the weight converges to -1 or 1 and 0 is an unstable point. And then, we will rigorously prove the global convergence of (3.5).

Given any $H > 0$, we define

$$S(H) = \left\{ w \,\middle|\, w^T Cw \leq H \right\}.$$

Lemma 3.3 *Given any $H \geq \bar{H}$, if $w(0) \in S(H)$ and $w(0) \notin V_\sigma^\perp$, then there exists constant $\theta_1 > 0$ and $\Pi_1 \geq 0$ such that*

$$\sum_{j=m+1}^{n} z_j^2(k) \leq \Pi_1 \cdot e^{-\theta_1 k},$$

for all $k \geq 0$, where

$$\theta_1 = ln\left(\frac{\xi\sigma + (1-\xi)H}{\xi\lambda_{m+1} + (1-\xi)H}\right)^2 > 0.$$

Proof: Since $w(0) \notin V_\sigma^\perp$, there must exist some $i(1 \leq i \leq m)$ such that $z_i(0) \neq 0$. Without loss of generality, assume that $z_1(0) \neq 0$.

From (3.6), it follows that

$$z_i(k+1) = \left[1 + \xi\left(\frac{\sigma}{w^T(k)Cw(k)} - 1\right)\right]z_i(k), 1 \leq i \leq m, \tag{3.8}$$

and

$$z_j(k+1) = \left[1 + \xi\left(\frac{\lambda_j}{w^T(k)Cw(k)} - 1\right)\right]z_j(k), m + 1 \leq j \leq n, \qquad (3.9)$$

for $k \geq 0$.

By Theorem 3.2, it follows that $w(k) \in S(H)$ for all $k \geq 0$. Given any $i(1 \leq i \leq n)$, it holds that

$$1 + \xi\left(\frac{\lambda_i}{w^T(k)Cw(k)} - 1\right) = 1 - \xi + \xi\left(\frac{\lambda_i}{w^T(k)Cw(k)}\right) \geq (1 - \xi) > 0,$$

for $k \geq 1$. Then, from (3.8) and (3.9), for each $j(m + 1 \leq j \leq n)$, it follows that

$$\left[\frac{z_j(k+1)}{z_1(k+1)}\right]^2 = \frac{\left[1 + \xi\left(\frac{\lambda_j}{w^T(k)Cw(k)} - 1\right)\right]^2}{\left[1 + \xi\left(\frac{\sigma}{w^T(k)Cw(k)} - 1\right)\right]} \cdot \left[\frac{z_j(k)}{z_1(k)}\right]^2$$

$$= \left[\frac{w^T(k)Cw(k) + \xi(\lambda_j - w^T(k)Cw(k))}{w^T(k)Cw(k) + \xi(\sigma - w^T(k)Cw(k))}\right]^2 \cdot \left[\frac{z_j(k)}{z_1(k)}\right]^2$$

$$\leq \left[\frac{\xi\lambda_j + (1 - \xi)H}{\xi\sigma + (1 - \xi)H}\right]^2 \cdot \left[\frac{z_j(k)}{z_1(k)}\right]^2$$

$$\leq \left[\frac{\xi\lambda_{m+1} + (1 - \xi)H}{\xi\sigma + (1 - \xi)H}\right]^2 \cdot \left[\frac{z_j(k)}{z_1(k)}\right]^2$$

$$= \left[\frac{z_j(0)}{z_1(0)}\right]^2 \cdot e^{-\theta_1(k+1)},$$

for all $k \geq 1$.

Since $w(k) \in S(H)$, $z_1(k)$ must be bounded; that is, there exists a constant $d > 0$ such that $z_1^2(k) \leq d$ for all $k \geq 0$. Then,

$$\sum_{j=m+1}^{n} z_j^2(k) = \sum_{j=m+1}^{n} \left[\frac{z_j(k)}{z_1(k)}\right]^2 \cdot z_1^2(k) \leq \Pi_1 e^{-\theta_1 k},$$

for $k \geq 0$, where

$$\Pi_1 = d \sum_{j=m+1}^{n} \left[\frac{z_j(0)}{z_1(0)}\right]^2 \geq 0.$$

This completes the proof.

Next, for convenience, denote

$$P(k) = \left[1 + \xi\left(\frac{\sigma}{w^T(k)Cw(k)} - 1\right)\right]^2 w^T(k)Cw(k), \qquad (3.10)$$

and

$$Q(k) = \sum_{i=m+1}^{n} \lambda_i \left[2(1 - \xi) + \frac{\xi(\lambda_i + \sigma)}{w^T(k)Cw(k)} \right] \left[\frac{\xi(\sigma - \lambda_i)}{w^T(k)Cw(k)} \right] z_i^2(k), \quad (3.11)$$

for all $k \geq 0$. Clearly, $P(k) \geq 0$ and $Q(k) \geq 0$ for all $k \geq 0$.

Lemma 3.4 *It holds that*

$$w^T(k+1)Cw(k+1) = P(k) - Q(k)$$

for $k \geq 0$.

Proof : From (1.8) and (3.6), it follows that

$$
\begin{aligned}
w^T(k+1)Cw(k+1) &= \sum_{i=1}^{n} \lambda_i \left[1 + \xi \left(\frac{\lambda_i}{w^T(k)Cw(k)} - 1 \right) \right]^2 z_i^2(k) \\
&= \sum_{i=1}^{n} \lambda_i \left[1 + \xi \left(\frac{\sigma}{w^T(k)Cw(k)} - 1 \right) \right]^2 z_i^2(k) \\
&\quad - \sum_{i=m+1}^{n} \lambda_i \left[2(1-\xi) + \frac{\xi(\lambda_i + \sigma)}{w^T(k)Cw(k)} \right] \left[\frac{\xi(\sigma - \lambda_i)}{w^T(k)Cw(k)} \right] z_i^2(k) \\
&= \left[1 + \xi \left(\frac{\sigma}{w^T(k)Cw(k)} - 1 \right) \right]^2 w^T(k)Cw(k) \\
&\quad - \sum_{i=m+1}^{n} \lambda_i \left[2(1-\xi) + \frac{\xi(\lambda_i + \sigma)}{w^T(k)Cw(k)} \right] \left[\frac{\xi(\sigma - \lambda_i)}{w^T(k)Cw(k)} \right] z_i^2(k) \\
&= P(k) - Q(k).
\end{aligned}
$$

The proof is complete.

Lemma 3.5 *If $w(0) \notin V_\sigma^\perp$, it holds that*

$$P(k - 1) \geq 4(1 - \xi)\xi\sigma,$$

for $k \geq 1$.

Proof : From (3.10), it follows that

$$
\begin{aligned}
P(k-1) &= \left[1 + \xi \left(\frac{\sigma}{w^T(k-1)Cw(k-1)} - 1 \right) \right]^2 w^T(k-1)Cw(k-1) \\
&\geq \min_{s>0} \left\{ \left[1 + \xi \left(\frac{\sigma}{s} - 1 \right) \right]^2 s \right\}
\end{aligned}
$$

for $k \geq 1$. By Lemma (3.1), it holds that

$$P(k - 1) \geq 4(1 - \xi)\xi\sigma, \quad k \geq 1.$$

The proof is complete.

Lemma 3.6 *There exists a positive constant $\Pi_2 > 0$ such that*

$$Q(k) \leq \Pi_2 \cdot e^{-\theta_1 k},$$

for $k \geq 0$.

Proof : From (3.11),

$$Q(k) = \sum_{i=m+1}^{n} \lambda_i \left[2(1 - \xi) + \frac{\xi(\lambda_i + \sigma)}{w^T(k)Cw(k)} \right] \left[\frac{\xi(\sigma - \lambda_i)}{w^T(k)Cw(k)} \right] z_i^2(k)$$

$$\leq \sum_{i=m+1}^{n} \lambda_i \left[2 + \frac{2\xi\sigma}{w^T(k)Cw(k)} \right] \left[\frac{\xi\sigma}{w^T(k)Cw(k)} \right] z_i^2(k)$$

$$\leq 2 \left[1 + \frac{\xi\sigma}{w^T(k)Cw(k)} \right] \left[\frac{\xi\sigma}{w^T(k)Cw(k)} \right] \cdot \sum_{i=m+1}^{n} \lambda_i z_i^2(k)$$

for $k \geq 0$. By Theorem 3.1 and Lemma 3.3,

$$Q(k) \leq 2 \left[1 + \frac{\xi\sigma}{4(1 - \xi)\xi\lambda_p} \right] \left[\frac{\xi\sigma}{4(1 - \xi)\xi\lambda_p} \right] \sum_{i=m+1}^{n} \lambda_i z_i^2(k)$$

$$\leq 2\Pi_1 \cdot \left[1 + \frac{\sigma}{4(1 - \xi)\lambda_p} \right] \left[\frac{\sigma}{4(1 - \xi)\lambda_p} \right] e^{-\theta_1 k}$$

$$\leq \Pi_2 e^{-\theta_1 k}$$

for $k \geq 0$, where

$$\Pi_2 = 2\Pi_1 \cdot \left[1 + \frac{\sigma}{4(1 - \xi)\lambda_p} \right] \left[\frac{\sigma}{4(1 - \xi)\lambda_p} \right].$$

The proof is complete.

Lemma 3.7 *Given any $H \geq \bar{H}$, if $w(0) \in S(H)$ and $w(0) \notin V_\sigma^\perp$, then there exists constants $\theta_2 > 0$ and $\Pi_5 > 0$ such that*

$$\left| \sigma - w^T(k + 1)Cw(k + 1) \right| \leq k \cdot \Pi_5 \cdot \left[e^{-\theta_2(k+1)} + \max \left\{ e^{-\theta_2 k}, e^{-\theta_1 k} \right\} \right],$$

for all $k \geq 1$, where

$$\begin{cases} \theta_2 = -\ln \delta, \\ \delta = \max \left\{ (1 - \xi)^2, \dfrac{\xi}{4(1 - \xi)} \right\}, \end{cases}$$

and $0 < \xi < 0.8$, $0 < \delta < 1$.

Proof: From (3.10), (3.11), and Lemma 3.4, it follows that

$$\sigma - w^T(k+1)Cw(k+1)$$
$$= \sigma - P(k) + Q(k)$$
$$= \sigma - \left[1 + \xi\left(\frac{\sigma}{w^T(k)Cw(k)} - 1\right)\right]^2 w^T(k)Cw(k) + Q(k)$$
$$= (\sigma - w^T(k)Cw(k))\left[(1-\xi)^2 - \frac{\xi^2\sigma}{w^T(k)Cw(k)}\right] + Q(k)$$
$$= (\sigma - w^T(k)Cw(k))\left[(1-\xi)^2 - \frac{\xi^2\sigma}{P(k-1) - Q(k-1)}\right] + Q(k)$$
$$= (\sigma - w^T(k)Cw(k))\left[(1-\xi)^2 - \frac{\xi^2\sigma}{P(k-1)}\right]$$
$$- \frac{(\sigma - w^T(k)Cw(k))\,\xi^2\sigma Q(k-1)}{P(k-1)\,(P(k-1) - Q(k-1))} + Q(k)$$
$$= (\sigma - w^T(k)Cw(k))\left[(1-\xi)^2 - \frac{\xi^2\sigma}{P(k-1)}\right]$$
$$- \frac{(\sigma - w^T(k)Cw(k))\,\xi^2\sigma}{P(k-1)w^T(k)Cw(k)}Q(k-1) + Q(k)$$
$$= (\sigma - w^T(k)Cw(k))\left[(1-\xi)^2 - \frac{\xi^2\sigma}{P(k-1)}\right]$$
$$- \xi^2\sigma\left[\frac{\sigma}{w^T(k)Cw(k)} - 1\right]\frac{Q(k-1)}{P(k-1)} + Q(k)$$

for $k \geq 1$.

Denote

$$V(k) = \left|\sigma - w^T(k)Cw(k)\right|,$$

for $k \geq 1$. It follows that

$$V(k+1) \;\leq\; V(k) \cdot \left|(1-\xi)^2 - \frac{\xi^2\sigma}{P(k-1)}\right|$$
$$+\xi^2\sigma\left[\frac{\sigma}{w^T(k)Cw(k)} + 1\right]\frac{Q(k-1)}{P(k-1)} + Q(k)$$
$$\leq\; \max\left\{(1-\xi)^2, \frac{\xi^2\sigma}{P(k-1)}\right\} \cdot V(k)$$
$$+\xi^2\sigma\left[\frac{\sigma}{w^T(k)Cw(k)} + 1\right]\frac{Q(k-1)}{P(k-1)} + Q(k)$$

for $k \geq 1$. By Lemma 3.5, it holds that

$$
\begin{aligned}
V(k+1) \leq{} & \max\left\{(1-\xi)^2, \frac{\xi}{4(1-\xi)}\right\} \cdot V(k) \\
& + \frac{\xi}{4(1-\xi)}\left[\frac{\sigma}{w^T(k)Cw(k)} + 1\right] Q(k-1) + Q(k),
\end{aligned}
$$

for $k \geq 1$. Denote

$$
\delta = \max\left\{(1-\xi)^2, \frac{\xi}{4(1-\xi)}\right\}.
$$

Clearly, if $0 < \xi < 0.8$, $0 < \delta < 1$. Then,

$$
V(k+1) \leq \delta \cdot V(k) + \frac{\xi}{4(1-\xi)}\left[\frac{\sigma}{w^T(k)Cw(k)} + 1\right] Q(k-1) + Q(k), k \geq 1.
$$

Denote

$$
\Pi_3 = \frac{\xi}{4(1-\xi)} \cdot \left[\frac{\sigma}{4(1-\xi)\xi\lambda_p} + 1\right].
$$

By Theorem 3.1 and Lemma 3.6,

$$
\begin{aligned}
V(k+1) & \leq \delta \cdot V(k) + \Pi_3 \cdot Q(k-1) + Q(k) \\
& \leq \delta \cdot V(k) + \Pi_4 \cdot e^{-\theta_1 k}
\end{aligned}
$$

where

$$
\Pi_4 = \Pi_2 \cdot \left(\Pi_3 e^{\theta_1} + 1\right).
$$

Then,

$$
\begin{aligned}
V(k+1) & \leq \delta^{k+1} V(0) + \Pi_4 \sum_{r=0}^{k} \delta^r e^{-\theta_1(k-r)} \\
& \leq \delta^{k+1} V(0) + \Pi_4 \sum_{r=0}^{k} \left(\delta e^{\theta_1}\right)^r e^{-\theta_1 k} \\
& \leq \delta^{k+1} V(0) + k \Pi_4 \cdot \max\left\{\delta^k, e^{-\theta_1 k}\right\} \\
& \leq k \cdot \Pi_5 \cdot \left[e^{-\theta_2(k+1)} + \max\left\{e^{-\theta_2 k}, e^{-\theta_1 k}\right\}\right],
\end{aligned}
$$

where $\theta_2 = -\ln\delta > 0$, and

$$
\Pi_5 = \max\left\{\left|\sigma - w^T(0)Cw(0)\right|, \Pi_4\right\} > 0.
$$

The proof is complete.

Lemma 3.8 *Suppose there exists constants $\theta > 0$ and $\Pi > 0$ such that*

$$
\frac{\xi}{w^T(k)Cw(k)} \left|\left(\sigma - w^T(k)Cw(k)\right) z_i(k)\right| \leq k \cdot \Pi e^{-\theta k}, (i = 1, \ldots, m)
$$

for $k \geq 1$. Then,

$$\lim_{k \to +\infty} z_i(k) = z_i^*, (i = 1, \ldots, m),$$

where $z_i^*(i = 1, \ldots, m)$ are constants.

Proof: Given any $\epsilon > 0$, there exists a $K \geq 1$ such that

$$\frac{\Pi K e^{-\theta K}}{(1 - e^{-\theta})^2} \leq \epsilon.$$

For any $k_1 > k_2 \geq K$, it follows that

$$\begin{aligned}
|z_i(k_1) - z_i(k_2)| &= \left| \sum_{r=k_2}^{k_1-1} [z_i(r+1) - z_i(r)] \right| \\
&\leq \sum_{r=k_2}^{k_1-1} \frac{\xi}{w^T(r)Cw(r)} \left| \left(\sigma - w^T(r)Cw(r)\right) z_i(r) \right| \\
&\leq \Pi \sum_{r=k_2}^{k_1-1} r e^{-\theta r} \\
&\leq \Pi \sum_{r=K}^{+\infty} r e^{-\theta r} \\
&\leq \Pi K e^{-\theta K} \cdot \sum_{r=0}^{+\infty} r \left(e^{-\theta} \right)^{r-1} \\
&= \frac{\Pi K e^{-\theta K}}{(1 - e^{-\theta})^2} \\
&\leq \epsilon, \quad (i = 1, \ldots, m).
\end{aligned}$$

This shows that each sequence $\{z_i(k)\}$ is a *Cauchy sequence*. By *Cauchy Convergence Principle*, there must exist constants $z_i^*(i = 1, \ldots, m)$ such that

$$\lim_{k \to +\infty} z_i(k) = z_i^*, \quad (i = 1, \ldots, m).$$

This completes the proof.

Theorem 3.3 *Suppose that $0 < \xi < 0.8$. if $w(0) \in S$ and $w(0) \notin V_\sigma^\perp$, then the trajectory of (3.5) starting from $w(0)$ will converge to a unit eigenvector associated with the largest eigenvalue of the covariance matrix C.*

Proof: By Lemma 3.3, there exists constants $\theta_1 > 0$ and $\Pi_1 \geq 0$ such that

$$\sum_{j=m+1}^{n} z_j^2(k) \leq \Pi_1 e^{-\theta_1 k},$$

for all $k \geq 0$. By Lemma 3.7, there exists constants $\theta_2 > 0$ and $\Pi_5 > 0$ such that

$$\left| \sigma - w^T(k+1)Cw(k+1) \right| \leq k \cdot \Pi_5 \cdot \left[e^{-\theta_2(k+1)} + \max \left\{ e^{-\theta_2 k}, e^{-\theta_1 k} \right\} \right]$$

for all $k \geq 0$.

Obviously, there exists constants $\theta > 0$ and $\Pi > 0$ such that

$$\frac{\xi}{w^T(k)Cw(k)} \left| \left(\sigma - w^T(k)Cw(k) \right) z_i(k) \right| \leq k \cdot \Pi e^{-\theta k}, (i = 1, \ldots, m)$$

for $k \geq 0$.

Using Lemma 3.3 and Lemma 3.8, it follows that

$$\begin{cases} \lim_{t \to +\infty} z_i(k) = z_i^*, (i = 1, \ldots, m) \\ \lim_{t \to +\infty} z_i(k) = 0, (i = m+1, \ldots, n). \end{cases}$$

Then,

$$\lim_{t \to +\infty} w(k) = \sum_{i=1}^{m} z_i^* v_i \in V_\sigma. \tag{3.12}$$

From (3.5), after the system becomes stable, it follows that

$$\lim_{k \to \infty} Cw(k) = \lim_{k \to \infty} w^T(k)Cw(k)w(k). \tag{3.13}$$

Substitute (3.12) into (3.13), we get

$$\sum_{i=1}^{m} \sigma z_i^* v_i = \sum_{i=1}^{m} \sigma(z_i^*)^2 \sum_{i=1}^{m} z_i^* v_i.$$

It is easy to see that

$$\sum_{i=1}^{m} (z_i^*)^2 = 1.$$

The proof is complete.

The above theorem shows almost all trajectories starting from $S(H)$ will converge to an eigenvector associated with the largest eigenvalue of the covariance matrix of C. $w(0)$ could be chosen arbitrarily except for the subspace $V_\sigma^\perp$. Clearly, the condition is easy to meet in practical applications.

3.5   Simulation and Discussion

Three simulation examples will be provided in this section to illustrate the proposed theory.

TABLE 3.1 The Norm of Initial Vector

0.41289	0.6404	1.8768	3.9806	6.241	9.2531

Example 1

Consider the simple example (3.7), Figure 3.2 gives the result that six trajectories of (3.7) starting from different points converge to the unit vector with precision $\epsilon = 0.00001$ and $\xi = 0.5$. The six initial vectors arbitrarily generated are in Table 3.1.

The left picture in Figure 3.2 presents the norm evolution of the weight. The learning rates evolution is shown in the right one, respectively. It is clear that learning rates converge to 0.5 and the evolution rates are also very fast. The Figure 3.3 shows the components of the weight with two different initial vectors converge to the unit vector, which is just an equilibrium point of (3.7). In the left picture in Figure 3.3, the initial vector

$$w(0) = \left[\ 0.03218, 0.020258, 0.0084304, 0.046076, 0.021929\ \right]^T$$

converges to the unit vector

$$w^* = \left[\ 0.5013, 0.31557, 0.13133, 0.71776, 0.34161\ \right]^T.$$

In the right one in Figure 3.3, the initial vector

$$w(0) = \left[\ 15.866, 21.48, 3.9579, 27.996, 19.861\ \right]^T$$

converges to the unit vector

$$w^* = \left[\ 0.36332, 0.49188, 0.090632, 0.64109, 0.45479\ \right]^T.$$

In fact, the initial vector may be arbitrary except for the origin. However, the figure can not clearly present it if the norm is very large or very small.

Example 2

The example 2 will illustrate the convergence of (3.5). First, consider a positive definitive matrix in R^2 that

$$C = \begin{bmatrix} 5 & 1 \\ 3 & 3 \end{bmatrix}.$$

The unit eigenvectors associated with the largest eigenvalue are $\pm[\sqrt{2}/2, \sqrt{2}/2]^T$, which are just equilibrium points of the system (3.5). Figure 3.4 shows the trajectories starting from the points randomly generated in R^2 converge to the equilibrium points with the precision $\epsilon = 0.00001$ and $\xi = 0.5$. From the results, it is apparent that all trajectories converge to the right direction associated with the largest eigenvalue of the matrix.

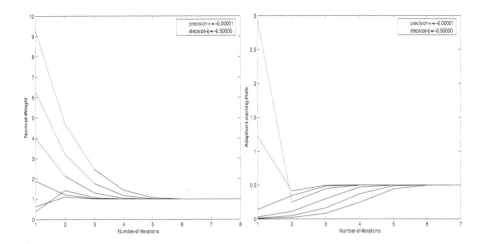

FIGURE 3.2
The weights converge to the unit one (left) and the learning rates converge to
0.5 (right) with six different initial values.

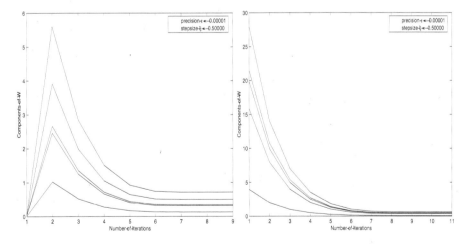

FIGURE 3.3
Convergence of (3.7) with the distinct initial vectors.

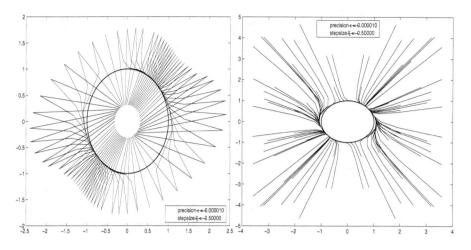

FIGURE 3.4
Global convergence of (3.5) with small initial vectors (left) and the large ones (right).

Next, consider a high-dimensional space. The covariance matrix is randomly generated as

$$C = \begin{bmatrix} 0.1712 & 0.1538 & 0.097645 & 0.036741 & 0.07963 & 0.12897 \\ 0.1538 & 0.13855 & 0.087349 & 0.033022 & 0.072609 & 0.11643 \\ 0.097645 & 0.087349 & 0.067461 & 0.032506 & 0.043641 & 0.070849 \\ 0.036741 & 0.033022 & 0.032506 & 0.019761 & 0.016771 & 0.025661 \\ 0.07963 & 0.072609 & 0.043641 & 0.016771 & 0.041089 & 0.062108 \\ 0.12897 & 0.11643 & 0.070849 & 0.025661 & 0.062108 & 0.098575 \end{bmatrix}.$$

We pick three different initial vectors. The components of the weight are updated independently. Figure 3.5, Figure 3.6, and Figure 3.7 give evolution results. The evolution of the weight is presented in the left pictures, and the evolution of learning rate is presented in right ones, respectively. In Figure 3.5, an initial vector

$$w(0) = \begin{bmatrix} 0.0092415, 0.087626, 0.028077, 0.062914, 0.085038, 0.0037035 \end{bmatrix}^T$$

converges to the vector

$$w^* = \begin{bmatrix} 0.57784, 0.52033, 0.33672, 0.13286, 0.27259, 0.43592 \end{bmatrix}^T.$$

Its learning rate converges to constant 0.98059. In Figure 3.6, the initial vector

$$w(0) = \begin{bmatrix} 0.37343, 0.20917, 0.51432, 0.068015, 0.49641, 0.21005 \end{bmatrix}^T$$

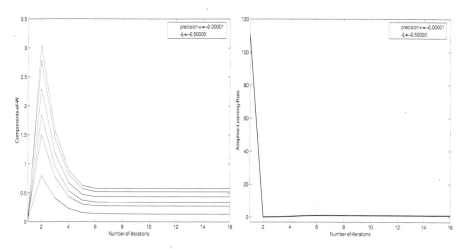

FIGURE 3.5
Convergence of (3.5) (left) and learning rate converge to 0.98059 (right).

converges to the vector

$$w^* = \left[\ 0.57784, 0.52033, 0.33672, 0.13285, 0.27259, 0.43592\ \right]^T.$$

Its learning rate converges to constant 0.98059. And then, the Figure 3.7 presents that the initial vector

$$w(0) = \left[\ 3340.5, 1380.1, 2416.5, 4045, 495.74, 4675.6\ \right]^T$$

converges to the vector

$$w^* = \left[\ 0.57784, 0.52033, 0.33671, 0.13286, 0.27259, 0.43592\ \right]^T.$$

Its learning rate converges to constant 0.98059.

Through a number of experiments, theory can be further confirmed that the system (3.5) is globally convergent and the learning rate approaches a constant, showing that the convergent rate achieved is very fast.

Example 3

By a numerical discretization procedure, the algorithm (3.5) can be written as (3.4). This algorithm (3.4) is more suitable for online learning with high performance. In this section, an image compression example will be provided to illustrate this aspect of the approach.

The Lenna picture in Figure 1.5 and the online observation sequence C_k (1.8) will be used to train the algorithm (3.4), replacing $C_{x(k)}$ by C_k with $\lim_{k \to \infty} C_k = C, (k > 0)$. The convergent precision ε can be determined easily according to the practical requirement. In the following simulations, let $\varepsilon = 0.00001$. As for the parameter ξ, the large ξ will speed up convergence, but it

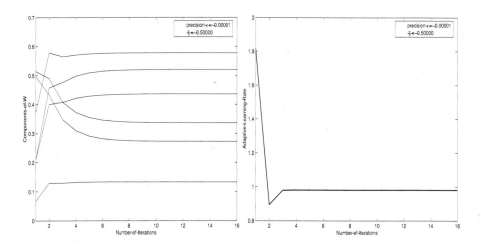

FIGURE 3.6
Convergence of (3.5) (left) and learning rate converge to 0.98059 (right).

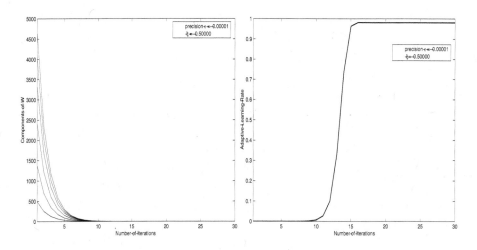

FIGURE 3.7
Convergence of (3.5) (left) and learning rate converge to 0.98059 (right).

also causes oscillation at the beginning phase of the training. A small ξ will make the evolution smoother with a relatively low convergent rate. Figure 3.8 shows that evolution trajectories of the same initial vector with the different ξ converge along the correct direction.

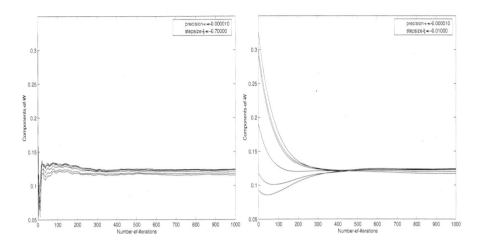

FIGURE 3.8
Convergence of (3.4) with $\xi = 0.7$ (left) and $\xi = 0.01$ (right).

In the following simulation, to observe the evolution trajectories clearly, let $\xi = 0.01$. Three distinct initial vectors are generated randomly. In Figure 3.9, Figure 3.10, Figure 3.11, the eight components trajectories in the weight vector are shown in left pictures and the right ones present the learning rate evolution, respectively. It can be observed that the weight vectors with different initial values fast converge to the principal component direction as the learning rates trend to a constant. Clearly, convergent rates derived are very fast.

Next, let $\xi = 0.5$, and we will further study the global convergence and convergent rate of (3.4). The eight arbitrarily initial values are selected apart from the origin, and each is repeatedly tested 20 times. The average results are shown in Table 3.2. It is clear that the arbitrary initial weight vector always converges along the correct direction with 100% success rate and the learning rate $\eta(k)$ converges to a constant such that the evolution rate derived does not become slower and slower. At the same time, the first principal component is extracted and the image is reconstructed. The average SNR has also been presented in Table 3.2. The reconstructed image is shown in Figure 3.12 with the initial value $2.3083e + 011$ and SNR 35.7643.

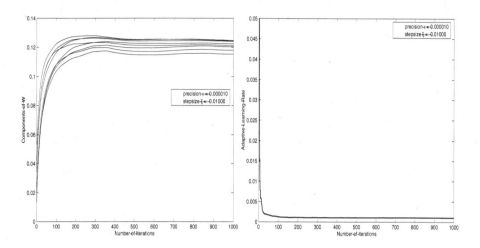

FIGURE 3.9
Convergence of (3.4) with with a small initial vector (norm = 0.2495).

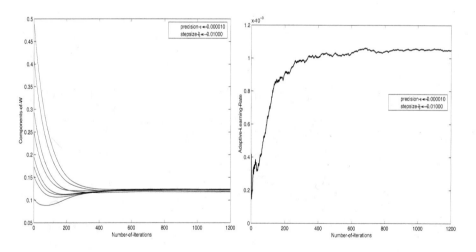

FIGURE 3.10
Convergence of (3.4) with a middle initial vector (norm = 2.5005).

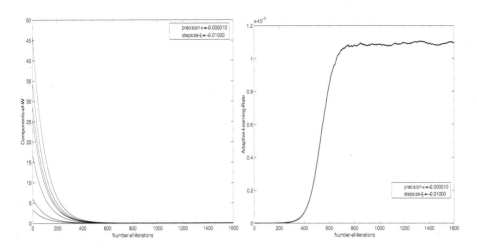

FIGURE 3.11
Convergence of (3.4) with a large initial vector (norm = 244.7259).

TABLE 3.2 The Average SNR

Ini.vaule	$\eta(k)$	SNR	Iter.Num.	Con.value	Suc.rate
2.0538e-008	0.0541	35.9727	997.5500	1.0000	100%
2.1899e-004	0.0535	36.1848	933.7500	1.0000	100%
0.2238	0.0532	35.7743	858.5500	1.0000	100%
4.3269	0.0531	35.8841	1.0529e+003	1.0000	100%
243.4699	0.0535	35.2856	992.1000	1.0000	100%
2.3382e+004	0.0535	35.8815	977.2500	1.0000	100%
2.3601e+006	0.0533	36.0921	961.2500	1.0000	100%
2.0855e+011	0.0537	35.9520	975.5500	1.0000	100%

FIGURE 3.12
The reconstructed image of Lenna image.

3.6   Conclusion

A non-zero-approaching adaptive learning rate is proposed to guarantee the global convergence of the Oja's algorithm. Using the DDT method, the global convergence of the corresponding DDT system is studied. Since the learning rate converges to a constant globally, learning procedure would become faster with time, and the choice of the learning parameter is simplified. Extensive experiments have confirmed that Oja's PCA algorithm with this non-approaching adaptive learning rate is suitable for online learning with high performance. Clearly, this non-zero-approaching adaptive learning rate could be generalized to other PCA algorithms.

4

GHA PCA Learning Algorithm

4.1 Introduction

Generalized Hebbian algorithm (GHA) is one of the most widely used algorithms in practical applications, introduced first in [156]. The algorithm is used to compute multiple principal component directions, which span a principal subspace.

In this chapter[2], non-zero-approaching learning rates will be proposed for the GHA. The proposed learning rates converge to some positive constants so that GHA with such learning rates can easily satisfy the computational and tracking requirements in applications. The non-zero-approaching learning rates make the convergence speed along all principal component directions approximately equal [131] so that the evolution is speeded up. In addition, the global convergence of the GHA can be guaranteed by using the non-zero-approaching learning rates.

This chapter is organized as follows. Section 2 describes the problem formulation and gives preliminaries. Convergence results are given in Section 3. Simulation results and discussions are presented in Section 4. Conclusions are drawn in Section 5.

4.2 Problem Formulation and Preliminaries

With the input sequence (1.1) and network model in Figure 1.2, let k be the time step and l the neuron index. The GHA [156] can be described by the

---

[2]Based on "Global convergence of GHA learning algorithm with nonzero-approaching learning rates," by(Jian Cheng Lv, Zhang Yi, and K. K. Tan), which appeared in IEEE Trans. Neural Networks, vol. 18, no. 6, pp. 1557–1571, Nov. 2007. ©[2007] IEEE.

following stochastic difference equations:

$$y_l(k) = w_l^T(k)x(k)$$
$$w_l(k+1) = w_l(k) + \eta_l(k) \times [y_l(k)x(k)$$
$$- \sum_{i=1}^{l-1} w_i(k)w_i(k)^T y_l(k)x(k) - y_l^2(k)w_l(k)],$$

(4.1)

where $\eta_l(k) > 0$ is the learning rate.

There are some problems relating to the convergence of (4.1). First, (4.1) may not converge if the learning rates are constants, a special case for $l = 1$ can be found in [195] and [210]. Second, if the learning rates approach zero, it is difficult to satisfy the computational and tracking requirements in practical applications [40, 195, 202, 210].

To overcome these problems mentioned above, this paper proposes to use the following non-zero-approaching adaptive learning rates for the GHA:

$$\eta_l(k) = \frac{\zeta}{y_l^2(k)}, \quad k > 0. \tag{4.2}$$

The non-zero-approaching adaptive learning rates not only allow the GHA to easily satisfy the computational and tracking requirements in applications, but they also guarantee the global convergence of the GHA. The GHA, with the learning rates (4.2), is written as

$$w_l(k+1) = w_l(k) + \frac{\zeta}{y_l^2(k)} \times [y_l(k)x(k)$$
$$- \sum_{i=1}^{l-1} w_i(k)w_i(k)^T y_l(k)x(k) - y_l^2(k)w_l(k)],$$

(4.3)

for $k > 0$, where $0 < \zeta < 1$.

In the following section, we study the convergence of the GHA with the proposed learning rate via DDT method.

Taking the conditional expectation $E\{w_l(k+1)/w_l(0), w_l(i), i < k\}$ on both sides of (4.2), there must exist a constant $\alpha > 0$ so that

$$E\left\{\frac{\zeta}{y_l^2(k)}\right\} = E\left\{\frac{\zeta}{w_l^T(k)x(k)x^T(k)w_l(k)}\right\}$$
$$= \alpha \cdot \frac{\zeta}{w_l^T(k)Cw_l(k)}$$
$$= \frac{\xi}{w_l^T(k)Cw_l(k)},$$

where $\xi = \alpha\zeta > 0$ and $C = E[x(k)x^T(k)]$. Denote

$$Y(k) = [y_l(k)x(k)$$
$$- \sum_{i=1}^{l-1} w_i(k)w_i(k)^T y_l(k)x(k) - y_l^2(k)w_l(k)].$$

Clearly, the probability $P(Y(k))$ and $P(\frac{\varsigma}{y_l^2(k)})$ only depend on the distribution of the variable X. It is easy to see that $P(Y(k))$ is not affected by the value of the learning rate (4.2); that is, the value of the learning rates (4.2) gives no information on the value of $Y(k)$. Thus, the learning rate (4.2) and $Y(k)$ are independent of each other. Taking the conditional expectation $E\{w_l(k+1)/w_l(0), w_l(i), i < k\}$ on both sides of (4.3), the corresponding DDT system for the GHA with learning rates (4.2) can be written as

$$
\begin{aligned}
w_l(k+1) &= w_l(k) + E\left\{\frac{\varsigma}{y_l^2(k)}\right\} \times E\{Y(k)\} \\
&= w_l(k) + \frac{\xi}{w_l(k)^T C w_l(k)} \times \\
&\quad E\left\{\left[x(k)x^T(k)w_l(k)\right.\right. \\
&\quad\left. - \sum_{i=1}^{l-1} w_i(k)w_i(k)^T x(k)x^T(k)w_l(k)\right. \\
&\quad\left.\left. - w_l^T(k)x(k)x^T(k)w_l(k)w_l(k)\right]\right\}, \\
&= w_l(k) + \frac{\xi}{w_l(k)^T C w_l(k)} \times \\
&\quad \left[Cw_l(k) - \sum_{i=1}^{l-1} w_i(k)w_i(k)^T C w_l(k)\right. \\
&\quad\left. - w_l^T(k)Cw_l(k)w_l(k)\right],
\end{aligned}
\tag{4.4}
$$

for $k > 0$, where $0 < \xi < 1$ and $C = E\{x(k)x^T(k)\}$.

The DDT system preserves the discrete time form of the original algorithm and gathers a more realistic behavior of the learning gain.

The covariance matrix C given in (4.4) is a symmetric nonnegative definite matrix. Let $\lambda_i (i = 1, \ldots, n)$ be all the eigenvalues of the matrix C ordered by $\lambda_1 > \lambda_2 > \ldots > \lambda_p > \lambda_{p+1} = \ldots = \lambda_n = 0$, and let v_i be the unit eigenvector of C associated with the eigenvalue λ_i. Then, $\{v_i | i = 1, 2, \ldots, n\}$ forms an orthonormal base of $\mathbb{R}^n$. Denote by V_{λ_i} the eigensubspace of the eigenvalue λ_i, and let $V_{\lambda_i}^\perp$ be the subspace perpendicular to V_{λ_i}. Denoting by V_0 the null subspace associated with the eigenvalue 0, clearly,

$$
V_0 = span\{v_{p+1}, \ldots, v_n\}.
$$

Since the vector set $\{v_i | i = 1, 2, \ldots n\}$ is an orthonormal basis of $\mathbb{R}^n$, for $k > 0$, $w_l(k) \in \mathbb{R}^n$ can be represented as

$$
w_l(k) = \sum_{i=1}^{n} z_i^{(l)}(k)v_i,
\tag{4.5}
$$

where $z_i^{(l)}(k)(i = 1, \ldots, n)$ are some coefficients and $1 \leq l \leq p$. It is noted that we require $1 \leq l \leq p$ since extracting eigenvectors associated with zero eigenvalue is uninteresting in practice. Clearly,

$$w_l^T(k)Cw_l(k) = \sum_{i=1}^{p} \lambda_i \left(z_i^{(l)}(k) \right)^2 \geq 0, \tag{4.6}$$

for all $k > 0$, where $1 \leq l \leq p$.

The following lemma will be used.

Lemma 4.1 *It holds that*

$$\left[1 - \xi + \frac{\xi D}{s} \right]^2 s \geq 4(1 - \xi)\xi D,$$

for $s > 0$ and $D > 0$, where $0 < \xi < 1$.

Refer to the Lemma 3.1 for the proof.

4.3 Convergence Analysis

In this section, we will study the convergence of (4.4). We will prove that starting from the initial vectors except those in the subspace $V_{\lambda_l}^{\perp}$, the algorithm (4.4) will converge to a vector in the subspace V_{λ_l}. Since the dimension of $V_{\lambda_l}^{\perp}$ is less than n, the measure of $V_{\lambda_l}^{\perp}$ is zero so that any unstable behavior cannot be observed in practical applications. From this point of view, the algorithm (4.4) is globally convergent. This proof will be completed using mathematic induction.

4.3.1 Outline of Proof

The mathematic induction will be used to prove that, by using algorithm (4.4), $w_l(k)$ converge to $\pm v_l$, $1 \leq l \leq p$.

For case $l = 1$, Lemmas 4.2 and 4.3 give the bounds of $\|w_1(k)\|$ as

$$0 < \sqrt{\frac{4(1-\xi)\xi\lambda_p}{\lambda_1}} \leq \|w_1(k+1)\| \leq \tilde{H}_1,$$

where $\tilde{H}_1$ is a constant. Then, Lemma 4.4 proves that $z_i^1, i = (2, \ldots, n)$ converge to zeroc and Lemma 4.5 proves that z_1^1 converge to a constant. Finally, the conclusion $w_1(k)$ converges to $\pm v_1$ is proved in Theorem 4.1.

Next, under the assumption that $\lim_{k \to \infty} w_i(k) = \pm v_i, (i = 1, \ldots, l-1)$, we prove that $w_l(k)$ converges to $\pm v_l$. Lemma 4.6 gives the bounds that

$$\sqrt{2\xi(1-\xi)\sqrt{\frac{\tilde{\Lambda}}{\lambda_1}}} < \|w_l(k+1)\| < \tilde{H}_l,$$

where $\tilde{H}_l$ is a constant. $\lim_{k \to \infty} z_i^{(l)}(k) = 0, (i = 1, \ldots, l-1)$ is proved in Lemma 4.7; $\lim_{k \to \infty} z_i^{(l)}(k) = 0, (i = l+1, \ldots, p)$ is proved in Lemma 4.9. Lemma 4.8 gives that z_l^l converges to a constant. Theorem 4.2 proves that $w_l(k)$ converges to $\pm v_l$.

The details are given next.

4.3.2 Case 1: $l = 1$

If $l = 1$, from (4.4), it follows that

$$
\begin{aligned}
w_1(k+1) &= w_1(k) + \frac{\xi}{w_1(k)^T C w_1(k)} \\
&\times \left[C w_1(k) - w_1^T(k) C w_1(k) w_1(k) \right],
\end{aligned}
$$

for $k > 0$, where $0 < \xi < 1$. Then, we have

$$
w_1(k+1) = (1 - \xi) w_1(k) + \xi \frac{C w_1(k)}{w_1(k)^T C w_1(k)}, \tag{4.7}
$$

for $k > 0$ and $0 < \xi < 1$. Substitute (4.5) and (4.6) into (4.7), we obtain

$$
\sum_{i=1}^{n} z_i^1(k+1) v_i = (1 - \xi) \sum_{i=1}^{n} z_i^1(k) v_i + \xi \frac{\sum_{i=1}^{n} z_i^1(k) C v_i}{w_1(k)^T C w_1(k)},
$$

for all $k > 0$. Multiply by v_i on both sides of the above equation, it follows that

$$
z_i^{(1)}(k+1) = (1 - \xi) z_i^{(1)}(k) + \xi \frac{\lambda_i z_i^{(1)}(k)}{w_1^T(k) C w_1(k)}, (i = 1, \ldots, n), \tag{4.8}
$$

for all $k > 0$.

We first prove that $\|w_1(k)\|$ is lower bounded.

Lemma 4.2 *Suppose that $w_1(0) \notin V_{\lambda_1}^{\perp}$. Then, it holds that*

$$
\|w_1(k+1)\| \geq \sqrt{\frac{4(1-\xi)\xi \lambda_p}{\lambda_1}} > 0,
$$

for all $k > 0$.

Proof: Since $\lambda_i \geq \lambda_p (i = 1, \ldots, p)$, it can be checked that

$$
\left[1 - \xi + \frac{\xi \lambda_i}{w_1^T(k) C w_1(k)} \right]^2 \geq \left[1 - \xi + \frac{\xi \lambda_p}{w_1^T(k) C w_1(k)} \right]^2
$$
$$
> 0, (1 \leq i \leq p),
$$

for $k > 0$. From (4.6) and (4.8), it follows that

$$
\begin{aligned}
& w_1^T(k+1)Cw_1(k+1) \\
=\ & \sum_{i=1}^{p} \lambda_i \left(z_i^{(1)}(k+1) \right)^2 \\
=\ & \sum_{i=1}^{p} \left[1 - \xi + \frac{\xi \lambda_i}{w_1^T(k)Cw_1(k)} \right]^2 \lambda_i \left(z_i^{(1)}(k) \right)^2 \\
\geq\ & \left[1 - \xi + \frac{\xi \lambda_p}{w_1^T(k)Cw_1(k)} \right]^2 \cdot w_1^T(k)Cw_1(k) \\
\geq\ & \min_{s>0} \left\{ \left[1 - \xi + \frac{\xi \lambda_p}{s} \right]^2 \cdot s \right\}
\end{aligned}
$$

for $k > 0$. Using Lemma 4.1, it follows that

$$
w_1^T(k+1)Cw_1(k+1) \geq 4(1-\xi)\xi \lambda_p,
$$

for $k > 0$. Since $\lambda_1 \|w_1(k+1)\|^2 \geq w_1^T(k+1)Cw_1(k+1)$ for $k \geq 0$, it follows that

$$
\|w_1(k+1)\| \geq \sqrt{\frac{4(1-\xi)\xi \lambda_p}{\lambda_1}} > 0,
$$

for $k > 0$. This completes the proof.

Next, we prove that $\|w_1(k)\|$ is upper bounded.

Lemma 4.3 *Suppose that $w_1(0) \notin V_{\lambda_1}^{\perp}$. There must exist a constant $\tilde{H}_1$ so that*

$$
\|w_1(k+1)\| < \tilde{H}_1,
$$

for all $k > 0$.

Proof: From (4.7), it follows that

$$
\|w_1(k+1)\| \leq (1-\xi)\|w_1(k)\| + \xi \frac{\|C\|\,\|w_1(k)\|}{w_1(k)^T Cw_1(k)}.
$$

Let $d_1 = \xi \dfrac{\|C\|}{\lambda_p}$. It follows that

$$
\begin{aligned}
\|w_1(k+1)\| \ &\leq\ (1-\xi)\|w_1(k)\| + \xi \frac{\|C\|\,\|w_1(k)\|}{\lambda_p \|w_1(k)\|^2} \\
&=\ (1-\xi)\|w_1(k)\| + \frac{d_1}{\|w_1(k)\|} \\
&=\ (1-\xi)^k \|w_1(0)\| + d_1 \sum_{r=1}^{k} \frac{(1-\xi)^{r-1}}{\|w_1(k-r)\|}
\end{aligned}
$$

for all $k > 0$.

Since $0 < \xi < 1$, by Lemma 4.2, it holds that

$$
\begin{aligned}
\|w_1(k+1)\| &\leq (1-\xi)^k \|w_1(0)\| \\
&\quad + d_1 \sqrt{\frac{\lambda_1}{4(1-\xi)\xi\lambda_p}} \cdot \sum_{r=1}^{k} (1-\xi)^{r-1} \\
&\leq \lim_{k \to \infty} \Big[(1-\xi)^k \|w_1(0)\| \\
&\quad + d_1 \sqrt{\frac{\lambda_1}{4(1-\xi)\xi\lambda_p}} \cdot \sum_{r=1}^{k} (1-\xi)^{r-1} \Big] \\
&= \tilde{H}_1,
\end{aligned}
$$

where

$$
\tilde{H}_1 = \frac{d_1}{\xi} \sqrt{\frac{\lambda_1}{4(1-\xi)\xi\lambda_p}}.
$$

Obviously, $\tilde{H}_1$ is a constant. The proof is completed.

From (4.8), it follows that

$$
z_1^{(1)}(k+1) = \left[1 - \xi + \frac{\xi\lambda_1}{w_1^T(k)Cw_1(k)} \right] z_1^{(1)}(k), \tag{4.9}
$$

and

$$
z_j^{(1)}(k+1) = \left[1 - \xi + \frac{\xi\lambda_j}{w_1^T(k)Cw_1(k)} \right] z_j^{(1)}(k), 2 \leq j \leq n, \tag{4.10}
$$

for $k > 0$. Then, given any $H_1 > 0$, we define

$$
S(H_1) = \{ w_1 | \, \| w_1 \| \leq H_1 \}.
$$

Lemma 4.4 *Suppose that $w_1(0) \notin V_{\lambda_1}^{\perp}$. Given any $H_1 \geq \tilde{H}_1$, if $w_1(0) \in S(H_1)$, then there exist constants $\theta_1 > 0$ and $\Pi_1 \geq 0$ such that*

$$
\sum_{j=2}^{n} (z_j^{(1)}(k))^2 \leq \Pi_1 \cdot e^{-\theta_1 k},
$$

for all $k > 0$, where

$$
\theta_1 = \ln \left(\frac{\xi\lambda_1 + (1-\xi)\lambda_1 H_1^2}{\xi\lambda_2 + (1-\xi)\lambda_1 H_1^2} \right)^2 > 0.
$$

Proof : Since $w_1(0) \notin V_{\lambda_1}^{\perp}$, then $z_1^{(1)}(0) \neq 0$. Given any $i(1 \leq i \leq n)$, it holds that

$$
1 - \xi + \xi \left(\frac{\lambda_i}{w_1^T(k)Cw_1(k)} \right) \geq (1-\xi) > 0,
$$

for $k > 0$, where $0 < \xi < 1$. Then, from (4.9) and (4.10), for each $j(2 \le j \le n)$, it follows that

$$
\left[\frac{z_j^{(1)}(k+1)}{z_1^{(1)}(k+1)} \right]^2
$$

$$
= \left[\frac{1 - \xi + \dfrac{\xi \lambda_j}{w_1^T(k)Cw_1(k)}}{1 - \xi + \dfrac{\xi \lambda_1}{w_1^T(k)Cw_1(k)}} \right]^2 \cdot \left[\frac{z_j^{(1)}(k)}{z_1^{(1)}(k)} \right]^2
$$

$$
= \left[\frac{(1-\xi)w_1^T(k)Cw_1(k) + \xi \lambda_j}{(1-\xi)w_1^T(k)Cw_1(k) + \xi \lambda_1} \right]^2 \cdot \left[\frac{z_j^{(1)}(k)}{z_1^{(1)}(k)} \right]^2
$$

$$
\le \left[\frac{\xi \lambda_j + (1-\xi)\lambda_1 H_1^2}{\xi \lambda_1 + (1-\xi)\lambda_1 H_1^2} \right]^2 \cdot \left[\frac{z_j^{(1)}(k)}{z_1^{(1)}(k)} \right]^2
$$

$$
\le \left[\frac{\xi \lambda_2 + (1-\xi)\lambda_1 H_1^2}{\xi \lambda_1 + (1-\xi)\lambda_1 H_1^2} \right]^2 \cdot \left[\frac{z_j^{(1)}(k)}{z_1^{(1)}(k)} \right]^2
$$

$$
= \left[\frac{z_j^{(1)}(0)}{z_1^{(1)}(0)} \right]^2 \cdot e^{-\theta_1(k+1)},
$$

for all $k > 0$.

Since $w_1(k) \in S(H_1)$, by Lemma 4.3, $z_1^{(1)}(k)$ must be bounded; that is, there exists a constant $b_1 > 0$ such that $(z_1^{(1)}(k))^2 \le b_1$ for all $k > 0$. Then,

$$
\sum_{j=2}^{n} (z_j^{(1)}(k))^2 = \sum_{j=2}^{n} \left[\frac{z_j^{(1)}(k)}{z_1^{(1)}(k)} \right]^2 \cdot (z_1^{(1)}(k))^2 \le \Pi_1 e^{-\theta_1 k},
$$

for $k > 0$, where

$$
\Pi_1 = b_1 \sum_{j=2}^{n} \left[\frac{z_j^{(1)}(0)}{z_1^{(1)}(0)} \right]^2 \ge 0.
$$

This completes the proof.

Lemma 4.5 *Suppose that $w_1(0) \notin V_{\lambda_1}^{\perp}$. Given any $H_1 \ge \tilde{H}_1$, if $w_1(0) \in S(H_1)$, then it holds that*

$$
\lim_{k \to +\infty} z_1^{(1)}(k) = (z_1^{(1)})^*,
$$

where $(z_1^{(1)})^*$ is a constant.

Proof : Since $w_1(0) \notin V_{\lambda_1}^{\perp}$, then $z_1^{(1)}(0) \neq 0$. From (4.9), it follows that

$$z_1^{(1)}(k+1) = (1-\xi)z_1^{(1)}(k) + \xi \left[\frac{\lambda_1 z_1^{(1)}(k)}{w_1^T(k)Cw_1(k)} \right],$$

for all $k > 0$, where $0 < \xi < 1$. From (4.9), $z_1^{(1)}(k) > 0$ for all $k \geq 0$ if $z_1^{(1)}(0) > 0$ and $z_1^{(1)}(k) < 0$ for all $k \geq 0$ if $z_1^{(1)}(0) < 0$. While $z_1^{(1)}(0) > 0$, by Lemma 4.2, there must exist a constant N_1 so that

$$0 < \frac{\lambda_1 z_1^{(1)}(k)}{w_1^T(k)Cw_1(k)} < N_1.$$

It follows that

$$
\begin{aligned}
& z_1^{(1)}(k+1) \\
\leq \quad & (1-\xi)z_1^{(1)}(k) + \xi N_1 \\
= \quad & (1-\xi)^k z_1^{(1)}(0) + \sum_{r=1}^{k}(1-\xi)^{r-1}\xi N_1 \\
= \quad & (1-\xi)^k z_1^{(1)}(0) + (1-(1-\xi)^k))N_1
\end{aligned}
$$

for all $k > 0$.

Since $0 < \xi < 1$, it holds that

$$\lim_{k \to \infty} \left[(1-\xi)^k z_1^{(1)}(0) + (1-(1-\xi)^k))N_1 \right] = N_1.$$

Thus, there must exist a constant $(z_1^{(1)})^*$ so that

$$\lim_{k \to +\infty} z_1^{(1)}(k+1) = (z_1^{(1)})^*.$$

Although $z_1^{(1)}(0) > 0$, the result can be obtained similarly. The proof is completed.

Theorem 4.1 *Suppose that $w_1(0) \notin V_{\lambda_1}^{\perp}$. If $w_1(0) \in S(H_1)$, then*

$$\lim_{k \to \infty} w_1(k) = \pm v_1.$$

Proof : By Lemma 4.4 and 4.5, it follows that

$$
\begin{cases}
\lim_{k \to \infty} z_1^{(1)}(k) = (z_1^{(1)})^*, \\
\lim_{k \to \infty} z_i^{(1)}(k) = 0, \quad (i = 2, \ldots, n).
\end{cases}
$$

Thus, we have

$$\lim_{k \to \infty} w_1(k) = (z_1^{(1)})^* v_1.$$

So, after the system (4.7) becomes stable, it follows that

$$\lim_{k \to \infty} C w_1(k) = \lim_{k \to \infty} w_1^T(k) C w_1(k).$$

Thus, it is easy to see that

$$\left[(z_1^{(1)})^* \right]^2 = 1.$$

The proof is completed.

According to the mathematical induction, next, we will prove the $w_l(k)$ converges to $\pm v_l$ under the assumptions $\lim_{k \to \infty} w_i(k) = \pm v_i, (i = 1, \ldots, l-1)$.

4.3.3   Case 2: $1 < l \le p$

Assume that there exists some constants $H_i (i = 1, \ldots, l-1)$ so that

$$\begin{cases} A1. & 0 < \|w_i(k)\| < H_i, (i = 1, \ldots, l-1) \\ A2. & \lim_{k \to \infty} w_i(k) = \pm v_i, (i = 1, \ldots, l-1) \\ A3. & w_l(k) \notin V_0, \text{ for all } k > 0. \end{cases}$$

We will prove $\lim_{k \to \infty} w_l(k) = \pm v_l$. According to the assumption $A2$, it follows that

$$w_i(k) = v_i + \varphi_i(k), \quad (i = 1, \ldots, l-1),$$

for all $k > 0$, where $\lim_{k \to \infty} \|\varphi_i(k)\| = 0$. From (4.4), we have

$$\begin{aligned} w_l(k+1) &= w_l(k) + \frac{\xi}{w_l^T(k) C w_l(k)} \Big[C w_l(k) \\ &\quad - \sum_{i=1}^{l-1} (v_i + \varphi_i(k))(v_i + \varphi_i(k))^T C w_l(k) \\ &\quad - w_l^T(k) C w_l(k) w_l(k) \Big], \\ &= w_l(k) + \frac{\xi}{w_l^T(k) C w_l(k)} \Big[C w_l(k) \\ &\quad - \sum_{i=1}^{l-1} v_i v_i^T C w_l(k) - w_l^T(k) C w_l(k) w_l(k) \Big] + \phi_l(k), \end{aligned} \tag{4.11}$$

for all $k > 0$, where

$$\phi_l(k) = -\xi \frac{\sum_{i=1}^{l-1} [\varphi_i(k) v_i^T + v_i \varphi_i(k)^T + \varphi_i(k)\varphi_i(k)^T] C w_l(k)}{w_l^T(k) C w_l(k)}.$$

Let $\phi_{li}(k) = v_i^T \phi_l(k)$. Substitute (4.5) and (4.6) into (4.11), for all $k > 0$, it follows that

$$z_i^{(l)}(k+1) = (1-\xi)z_i^{(l)}(k) + \phi_{li}(k),, \qquad (4.12)$$

for $i = 1, \ldots, l-1$, and

$$z_i^{(l)}(k+1) = \left\{ 1 - \xi + \frac{\xi\lambda_i}{w_l^T(k)Cw_l(k)} \right\} z_i^{(l)}(k) + \phi_{li}(k), \qquad (4.13)$$

for $i = l, \ldots, n$, where $\phi_{li}(k) = v_i^T \phi_l(k)$.

Let

$$\tilde{C}_l(k) = C - \sum_{i=1}^{l-1} w_i(k)w_i^T(k)C, \quad (1 \le l \le p),$$

for all $k > 0$. According to the assumption $A1$, there must exist a constant M_l so that $0 \le \|\tilde{C}_l(k)\| < M_l$ for all $k \ge 0$. Denote by $V_0^{(l)}(k)$, the null subspace associated with the eigenvalue 0 of the matrix $\tilde{C}_l(k)$ and $\Lambda_{min}^{(l)}(k)$ the smallest nonzero eigenvalue of $\tilde{C}_l(k)^T \tilde{C}_l(k)$. Clearly, $\Lambda_{min}^{(l)}(k) > 0$ for all $k > 0$. From (4.4), it follows that

$$w_l(k+1) = (1-\xi)w_l(k) + \xi\frac{\tilde{C}_l(k)w_l(k)}{w_l(k)^T Cw_l(k)}, \qquad (4.14)$$

for $k > 0$ and $0 < \xi < 1$.

Lemma 4.6 *Suppose that $w_l(k) \notin V_0^{(l)}(k)$ for all $k > 0$. $\|w_l(k+1)\|$ is ultimately bounded; that is, constants $K_l > 0$ must exist so that*

$$\sqrt{2\xi(1-\xi)\sqrt{\frac{\tilde{\Lambda}}{\lambda_1}}} < \|w_l(k+1)\| < \tilde{H}_l,$$

for all $k \ge K_l$, where $\tilde{\Lambda} = \min\{\Lambda_{min}^{(l)}(k), k \ge 0\}$ and $\tilde{H}_l = \dfrac{M_l}{\lambda_p\sqrt{2\xi(1-\xi)\sqrt{\frac{\tilde{\Lambda}}{\lambda_1}}}}$.

Proof : From (4.14), it follows that

$$
\begin{aligned}
\|w_l(k+1)\|^2 &= w_l(k+1)^T w_l(k+1) \\
&= (1-\xi)^2\|w_l(k)\|^2 \\
&\quad + 2\xi(1-\xi)\frac{w_l(k)^T \tilde{C}_l(k)^T w_l(k)}{w_l(k)^T Cw_l(k)} \\
&\quad + \xi^2 \frac{w_l(k)^T \tilde{C}_l(k)^T \tilde{C}_l(k)w_l(k)}{[w_l(k)^T Cw_l(k)]^2},
\end{aligned}
$$

for $k > 0$. Since $w_l(k) \notin V_0^{(l)}(k)$ for all $k \ge 0$ and the assumption $A2$, there must exist a constant K_l such that

$$w_l(k)^T \tilde{C}_l(k)w_l(k) \ge 0,$$

for all $k > K_l$. It follows that

$$
\begin{aligned}
\|w_l(k+1)\|^2 &\geq (1-\xi)^2 \|w_l(k)\|^2 \\
&\quad + \xi^2 \frac{w_l(k)^T \tilde{C}_l(k)^T \tilde{C}_l(k) w_l(k)}{[w_l(k)^T C w_l(k)]^2} \\
&= (1-\xi)^2 \|w_l(k)\|^2 + \xi^2 \frac{\tilde{\Lambda}}{\lambda_1 \|w_l(k)\|^2} \\
&\geq 2\xi(1-\xi) \sqrt{\frac{\tilde{\Lambda}}{\lambda_1}}
\end{aligned}
$$

for $k > K_l$, where $\tilde{\Lambda} = \min\{\Lambda^{(l)}_{min}(k), k > 0\}$.

Since $0 < \xi < 1$, from the assumption $A3$ and (4.14), it follows that

$$
\begin{aligned}
\|w_l(k+1)\| &\leq (1-\xi)\|w_l(k)\| + \xi \frac{M_l}{\lambda_p \|w_l(k)\|} \\
&= (1-\xi)^k \|w_l(0)\| + \xi \frac{M_l}{\lambda_p} \sum_{i=1}^{k} \frac{(1-\xi)^{i-1}}{\|w_1(k-i)\|} \\
&\leq (1-\xi)^k \|w_l(0)\| + \xi \tilde{H}_l \sum_{i=1}^{k} (1-\xi)^{i-1} \\
&\leq \tilde{H}_l,
\end{aligned}
$$

for $k \geq K_l$, where $\tilde{H}_l = \dfrac{M_l}{\lambda_p \sqrt{2\xi(1-\xi)\sqrt{\frac{\tilde{\Lambda}}{\lambda_1}}}}$. The proof is completed.

This lemma requires $w_l(k) \notin V_0^{(l)}(k)$ for all $k \geq 0$. This condition can be usually satisfied. It is because $\tilde{C}_l(k)$ will change with time and $V_0^{(l)}(k)$ is only a varying subspace. However, $l \leq p$ so that $\tilde{C}_l(k)$ may not converge to zero with time.

By Lemma 4.6, we have

$$\|\phi_l(k)\|$$

$$= \left\| \xi \frac{\sum_{i=1}^{l-1}[\varphi_i(k)v_i^T + v_i\varphi_i(k)^T + \varphi_i(k)\varphi_i(k)^T]Cw_l(k)}{w_l^T(k)Cw_l(k)} \right\|$$

$$\leq \left\| \xi \frac{\sum_{i=1}^{l-1}[\varphi_i(k)v_i^T + v_i\varphi_i(k)^T + \varphi_i(k)\varphi_i(k)^T]Cw_l(k)}{w_l^T(k)Cw_l(k)} \right\|$$

$$\leq \xi \frac{\lambda_1}{\lambda_p \sqrt{2\xi(1-\xi)}\sqrt{\frac{\bar{\Lambda}}{\lambda_1}}} \times$$

$$\left\| \xi \sum_{i=1}^{l-1}[\varphi_i(k)v_i^T + v_i\varphi_i(k)^T + \varphi_i(k)\varphi_i(k)^T] \right\|,$$

for $k > K_l$. By assumption $A2$, $\lim_{k\to\infty} \|\phi_l(k)\| = 0$. Clearly, $\lim_{k\to\infty} \phi_{li}(k) = 0$.

Lemma 4.7 *Suppose that* $\lim_{k\to\infty} \phi_{li}(k) = 0, (i = 1, \ldots, l-1)$. *It holds that*

$$\lim_{k\to\infty} z_i^{(l)}(k) = 0, (i = 1, \ldots, l-1).$$

Proof: From (4.12), given any $i(i = 1, \ldots, l-1)$, it follows that

$$z_i^{(l)}(k+1) = (1-\xi)\, z_i^{(l)}(k) + \phi_{li}(k),$$

$$= (1-\xi)^k z_i^{(l)}(0) + \sum_{r=1}^{k}(1-\xi)^{r-1}\phi_{li}(k-r)$$

for all $k > 0$. Since $0 < \xi < 1$ and $\lim_{k\to\infty} \phi_{li}(k) = 0$, it holds that

$$\lim_{k\to\infty} z_i^{(l)}(k+1)$$

$$= \lim_{k\to\infty}\left[(1-\xi)^k z_i^{(l)}(0) + \sum_{r=1}^{k}(1-\xi)^{r-1}\phi_{li}(k-r)\right]$$

$$= 0.$$

The proof is completed.

Given any $H_l > 0$, we define

$$S(H_l) = \{w_l| \parallel w_l \parallel \leq H_l\}.$$

Lemma 4.8 *Suppose* $\lim\limits_{k\to\infty} \phi_{li}(k) = 0, (i = l, \ldots, p)$. *Given any* $H_l \geq \tilde{H}_l$, *if* $w_l(0) \in S(H_l)$, *there must exist a constant* $\delta > 0$ *so that*

$$\lim_{k\to+\infty} z_l^{(l)}(k) = (z_l^{(l)})^*,$$

and $|(z_l^{(l)})^*| > \delta$, *where* $(z_l^{(l)})^*$ *is a constant.*

Proof : From (4.13), if $i = l$, it follows that

$$z_l^{(l)}(k+1) = \left\{1 - \xi + \frac{\xi\lambda_l}{w_l^T(k)Cw_l(k)}\right\} z_l^{(l)} + \phi_{ll}(k),$$

for all $k > 0$. Since $\lim\limits_{k\to\infty} \phi_{li}(k) = 0, (i = l, \ldots, p)$, there exists a $\tilde{K}_l$ so that

$$\begin{cases} z_l^{(l)}(k+1) > 0, & \text{if } z_l^{(l)}(\tilde{K}_l) > 0 \text{ for all } k \geq \tilde{K}_l, \\ z_l^{(l)}(k+1) < 0, & \text{if } z_l^{(l)}(\tilde{K}_l) < 0 \text{ for all } k \geq \tilde{K}_l. \end{cases}$$

Since $w_l(0) \in S(H_l)$, by Lemma 4.6, $w_l(k) \in S(H_l)$ for all $k \geq K_l$. While $z_l^{(l)}(\tilde{K}_l) > 0$, these exists a constant N_l so that

$$\left\{\frac{\xi\lambda_l}{w_l^T(k)Cw_l(k)}\right\} z_l^{(l)}(k) + \phi_{ll}(k) < N_l,$$

for all $k \geq \max\{K_l, \tilde{K}_l\}$. It follows that

$$\begin{aligned} z_l^{(l)}(k+1) & \\ \leq \ & (1-\xi)z_l^{(l)}(k) + N_l \\ = \ & (1-\xi)^k z_l^{(l)}(0) + \sum_{r=1}^{k}(1-\xi)^{r-1} N_l \\ = \ & (1-\xi)^k z_l^{(l)}(0) + (1-(1-\xi)^k) N_l, \end{aligned}$$

for all $k \geq K_l$. Since $0 < \xi < 1$, it holds that

$$\lim_{k\to\infty}\left[(1-\xi)^k |z_l^{(l)}(0)| + (1-(1-\xi)^k) N_l\right] = N_l.$$

Clearly, there must exist a constant $(z_l^{(l)})^*$ so that

$$\lim_{k\to+\infty} z_l^{(l)}(k) = (z_l^{(l)})^*,$$

where $(z_l^{(l)})^*$ is a constant. While $z_l^{(l)}(\tilde{K}_l) > 0$, the same result can be obtained similarly.

Next, we will prove that there exits a constant $\delta > 0$ so that $|(z_l^{(l)})^*| > \delta$.

If this is not true, then it means $\lim_{k \to +\infty} |z_l^{(l)}(k)| = 0$. We will prove that it leads to a contradiction.

Since $\lim_{k \to \infty} \phi_{li}(k) = 0, (i = l, \ldots, p)$, from (4.13), it follows that

$$
\begin{aligned}
& |z_l^{(l)}(k+1)| \\
= \; & \left| \left[(1 - \xi) + \frac{\xi \lambda_l}{w_l^T(k)Cw_l(k)} \right] z_l^{(l)}(k) + \phi_{ll}(k) \right|, \\
= \; & \left| (1 - \xi) + \frac{\xi \lambda_l}{w_l^T(k)Cw_l(k)} \right| |z_l^{(l)}(k)| + \phi_{ll}'(k),
\end{aligned}
$$

for all $k \geq K_l$, where $\lim_{k \to +\infty} \phi_{ll}'(k) = 0$. And for $i(i = l+1, \ldots, n)$, we have

$$
\begin{aligned}
& |z_i^{(l)}(k+1)| \\
= \; & \left| \left[(1 - \xi) + \frac{\xi \lambda_i}{w_l^T(k)Cw_l(k)} \right] z_i^{(l)}(k) + \phi_{li}(k) \right|, \\
= \; & \left| (1 - \xi) + \frac{\xi \lambda_i}{w_l^T(k)Cw_l(k)} \right| |z_i^{(l)}(k)| + \phi_{li}'(k),
\end{aligned}
$$

for all $k \geq K_l$, where $\lim_{k \to +\infty} \phi_{ll}'(k) = 0$. Clearly,

$$
\begin{aligned}
(1 - \xi) + \frac{\xi \lambda_l}{w_l^T(k)Cw_l(k)} \; & \geq \; (1 - \xi) + \frac{\xi \lambda_i}{w_l^T(k)Cw_l(k)} \\
& > \; 0, (i = l+1, \ldots, n).
\end{aligned}
$$

If k is sufficiently large, it is not difficult to show that $|z_l^{(l)}(k+1)| \geq |z_i^{(l)}(k+1)|$. Thus, if $\lim_{k \to +\infty} |z_l^{(l)}(k)| = 0$, then $\lim_{k \to +\infty} |z_i^{(l)}(k)| = 0 (i = l+1, \ldots, n)$. So,

$$
\lim_{k \to \infty} \sum_{i=l}^{p} \lambda_i (z_i^{(l)}(k))^2 = 0.
$$

However, from (4.7), we have

$$
w_l(k)^T Cw_l(k) = \sum_{i=1}^{l-1} \lambda_i (z_i^{(l)}(k))^2 + \sum_{i=l}^{p} \lambda_i (z_i^{(l)}(k))^2,
$$

for all $k \geq 0$. From Lemmas 4.6 and 4.7, it holds that

$$
\lim_{k \to \infty} \sum_{i=l}^{p} \lambda_i (z_i^{(l)}(k))^2 > 2 \lambda_p \xi (1 - \xi) \sqrt{\frac{\tilde{\Lambda}}{\lambda_1}} > 0,
$$

which is a contradiction. This show that there must exit a constant $\delta > 0$ so that $|(z_l^{(l)})^*| > \delta$.

This completes the proof.

Lemma 4.9 *Suppose that* $\lim\limits_{k\to\infty} |\phi_{li}| = 0, (i = l, \ldots, p)$ *and* $w_l(0) \notin V_{\lambda_l}^{\perp}$. *Given any* $H_l \geq \tilde{H}_l$, *if* $w_l(0) \in S(H_l)$, *then there exists constant* $\theta_l > 0$, $\Pi_l \geq 0$, *and* $\tilde{K}_l > 0$ *such that*

$$\sum_{j=l+1}^{n} (z_j^{(l)}(k))^2 \leq \Pi_l \cdot e^{-\theta_l k},$$

for all $k > \max\{K_l, \tilde{K}_l\}$, *where*

$$\theta_l = ln \left(\frac{\xi\lambda_l + (1-\xi)\lambda_l H_l^2}{\xi\lambda_{l+1} + (1-\xi)\lambda_l H_l^2} \right)^2 > 0.$$

Proof : Since $w_l(0) \notin V_{\lambda_l}^{\perp}$, then $z_l^{(l)}(0) \neq 0$. Then, from (4.13), for each $j(l+1 \leq j \leq n)$, it follows that

$$\left[\frac{z_j^{(l)}(k+1)}{z_l^{(l)}(k+1)} \right]^2$$

$$= \left[\frac{\left(1 - \xi + \dfrac{\xi\lambda_j}{w_l^T(k)Cw_l(k)} \right) z_j^{(l)}(k) + \phi_{lj}}{\left(1 - \xi + \dfrac{\xi\lambda_l}{w_l^T(k)Cw_l(k)} \right) z_l^{(l)}(k) + \phi_{ll}} \right]^2$$

$$= \left[\frac{1 - \xi + \dfrac{\xi\lambda_j}{w_l^T(k)Cw_l(k)}}{1 - \xi + \dfrac{\xi\lambda_l}{w_l^T(k)Cw_l(k)}} \right]^2 \cdot \left[\frac{z_j^{(l)}(k)}{z_l^{(l)}(k)} \right]^2 + \psi_l(k)$$

for all $k > 0$, where

$$\psi_l(k)$$
$$= \left[2\left(1 - \xi + \frac{\xi\lambda_l}{w_l^T(k)Cw_l(k)} \right) \right.$$
$$\times \left(1 - \xi + \frac{\xi\lambda_j}{w_l^T(k)Cw_l(k)} \right) z_j^{(l)}(k)z_l^{(l)}(k)$$
$$+ \left(1 - \xi + \frac{\xi\lambda_l}{w_l^T(k)Cw_l(k)} \right) z_l^{(l)}(k)\phi_{lj}$$
$$+ \left. \left(1 - \xi + \frac{\xi\lambda_j}{w_l^T(k)Cw_l(k)} \right) z_j^{(l)}(k)\phi_{ll} \right] \cdot \left[\frac{P_l(k)}{Q_l(k)^2} \right].$$

Meanwhile,

$$P_l(k) = \left[\left(1 - \xi + \frac{\xi\lambda_l}{w_l^T(k)Cw_l(k)} \right) z_l^{(l)}(k) \right] \phi_{lj}$$
$$- \left[\left(1 - \xi + \frac{\xi\lambda_j}{w_l^T(k)Cw_l(k)} \right) z_j^{(l)}(k) \right] \phi_{ll},$$

and

$$
Q_l(k) = \left\{ \left[\left(1 - \xi + \frac{\xi \lambda_l}{w_l^T(k)Cw_l(k)} \right) z_l^{(l)}(k) \right]^2 \right.
$$
$$
\left. + \left[\left(1 - \xi + \frac{.\xi \lambda_l}{w_l^T(k)Cw_l(k)} \right) z_l^{(l)}(k) \right]^2 \phi_{ll} \right\},
$$

for all $k > 0$. Since $\lim_{k \to \infty} |\phi_{li}| = 0, (i = l, \ldots, p)$, by Lemma 4.6, it follows that $\lim_{k \to \infty} |P_l(k)| = 0$. By Lemma 4.8, $\lim_{k \to \infty} |Q_l(k)|^2 > \delta^2$. So, it is easy to see that $\lim_{k \to \infty} |\psi_l(k)| = 0$.

Given any $i(1 \le i \le n)$, it can be checked that

$$
1 - \xi + \xi \left(\frac{\lambda_i}{w_l^T(k)Cw_l(k)} \right) \ge (1 - \xi) > 0,
$$

for $k \ge 1$, where $0 < \xi < 1$. Thus, there must exist a constant $\tilde{K}_l$ so that

$$
\left[\frac{z_j^{(l)}(k+1)}{z_l^{(l)}(k+1)} \right]^2 \le \left[\frac{\xi \lambda_j + (1-\xi)\lambda_l H_l^2}{\xi \lambda_l + (1-\xi)\lambda_l H_l^2} \right]^2 \cdot \left[\frac{z_j^{(l)}(k)}{z_l^{(l)}(k)} \right]^2
$$
$$
\le \left[\frac{\xi \lambda_{l+1} + (1-\xi)\lambda_l H_l^2}{\xi \lambda_l + (1-\xi)\lambda_l H_l^2} \right]^2 \cdot \left[\frac{z_j^{(l)}(k)}{z_l^{(l)}(k)} \right]^2
$$
$$
= \left[\frac{z_j^{(l)}(0)}{z_l^{(l)}(0)} \right]^2 \cdot e^{-\theta_l(k+1)},
$$

for $k > \tilde{K}_l$.

Since $w_l(0) \in S(H_l)$, by Lemma 4.6, $z_l^{(l)}(k)$ must be ultimately bounded; that is, there exists a constant $b_l > 0$ such that $(z_l^{(l)}(k))^2 \le b_l$ for all $k \ge K_l$. Then,

$$
\sum_{j=l+1}^{n} (z_j^{(l)}(k))^2 = \sum_{j=l+1}^{n} \left[\frac{z_j^{(l)}(k)}{z_l^{(l)}(k)} \right]^2 \cdot (z_l^{(l)}(k))^2 \le \Pi_l e^{-\theta_l k},
$$

for $k \ge \max\{K, \tilde{K}_l\}$, where

$$
\Pi_l = b_l \sum_{j=l+1}^{n} \left[\frac{z_j^{(l)}(0)}{z_l^{(l)}(0)} \right]^2 \ge 0.
$$

This completes the proof.

Theorem 4.2 *Suppose that* $\lim_{k \to \infty} |\phi_{li}| = 0, (i = 1, \ldots, p)$ *and* $w_l(0) \notin V_{\lambda_l}^\perp$. *Given any* $H_l \ge \tilde{H}_l$, *if* $w_l(0) \in S(H_l)$, *then*

$$
\lim_{k \to \infty} w_l(k) = \pm v_l,
$$

where $1 \leq l \leq p$.

Proof : By Lemmas 4.7, 4.8, and 4.9, it follows that

$$
\begin{cases}
\lim\limits_{t \to +\infty} z_i^{(l)}(k) = 0, (i = 1, \ldots, l-1) \\
\lim\limits_{t \to +\infty} z_l^{(l)}(k) = (z_l^{(l)})^*, \\
\lim\limits_{t \to +\infty} z_i^{(l)}(k) = 0, (i = l+1, \ldots, n).
\end{cases}
$$

Then,

$$
\lim_{k \to +\infty} w_l(k) = (z_l^{(l)})^* v_l. \tag{4.15}
$$

From (4.4), after the system becomes stable, it follows that

$$
\begin{aligned}
&\lim_{k \to \infty} C w_l(k) \\
&= \lim_{k \to \infty} \left[\sum_{i=1}^{l-1} v_i(k) v_i(k)^T C w_l(k) + w_l^T(k) C w_l(k) w_l(k) \right].
\end{aligned} \tag{4.16}
$$

Substitute (4.15) into (4.16), we get

$$
\lambda_l (z_l^{(l)})^* v_l = \lambda_l ((z_l^{(l)})^*)^2 (z_l^{(l)})^* v_l.
$$

It is easy to see that

$$
\left((z_l^{(l)})^* \right)^2 = 1.
$$

The proof is completed.

In the previous theorems, the condition $w_l(0) \notin V_{\lambda_l}^\perp$ is required. Since any small disturbance can result in $w_l(0) \notin V_{\lambda_l}^\perp$, it is easy to meet the condition in practical applications.

In our assumptions, the assumption $w_l(k) \notin V_0$ for all $k \geq 0$ is also required. Since extracted principal directions are associated with the nonzero eigenvalues and the evolution trajectories are always close to the principal directions, the condition is satisfied usually. The following lemma will further show this point.

Lemma 4.10 *Suppose that* $\lim\limits_{k \to \infty} \phi_{li}(k) = 0, (i = 1, \ldots, p)$. *Then, there exists a constant* $\hat{K}_l$ *so that*

$$
0 < \sqrt{\frac{4(1-\xi)\xi\lambda_p}{\lambda_1}} \leq \|w_l(k+1)\| \leq \frac{M_l}{\lambda_p} \sqrt{\frac{\lambda_1}{4(1-\xi)\xi\lambda_p}},
$$

for all $k \geq \max\{K_l, \hat{K}_l\}$, *where* $0 < \xi < 1$.

Proof : It can be checked that

$$\left[1 - \xi + \frac{\xi\lambda_i}{w_l^T(k)Cw_l(k)}\right]^2 \geq \left[1 - \xi + \frac{\xi\lambda_p}{w_l^T(k)Cw_l(k)}\right]^2$$
$$> 0,$$

for $k > 0$. From (4.6), (4.12), and (4.13), the following Γ is obtained,

$$\Gamma = w_l^T(k+1)Cw_l(k+1)$$

$$= \sum_{i=1}^{p} \lambda_i \left(z_i^{(l)}(k+1)\right)^2$$

$$= \sum_{i=1}^{l-1} \left[(1-\xi)z_i^{(l)}(k) + \phi_{li}(k)\right]^2 \lambda_i + \sum_{i=l}^{p} \left\{\left[1 - \xi + \frac{\xi\lambda_i}{w_l^T(k)Cw_l(k)}\right]z_i^{(l)}(k) + \phi_{li}(k)\right\}^2 \lambda_i$$

$$\geq \sum_{i=l}^{p} \left\{\left[1 - \xi + \frac{\xi\lambda_i}{w_l^T(k)Cw_l(k)}\right]z_i^{(l)}(k) + \phi_{li}(k)\right\}^2 \lambda_i$$

$$= \sum_{i=l}^{p} \left\{\left[1 - \xi + \frac{\xi\lambda_i}{\displaystyle\sum_{j=1}^{l-1}\lambda_j\left(z_j^{(l)}(k)\right)^2 + \sum_{j=l}^{p}\lambda_j\left(z_j^{(l)}(k)\right)^2}\right]z_i^{(l)}(k) + \phi_{li}(k)\right\}^2 \lambda_i$$

$$= \sum_{i=l}^{p} \left\{\left[1 - \xi + \frac{\xi\lambda_i}{\displaystyle\sum_{j=l}^{p}\lambda_j\left(z_j^{(l)}(k)\right)^2}\right]z_i^{(l)}(k) + \phi_{li}(k)\right.$$

$$\left. - \frac{\xi\lambda_i z_i^{(l)}(k)\displaystyle\sum_{j=1}^{l-1}\lambda_j\left(z_j^{(l)}(k)\right)^2}{w_l^T(k)Cw_l(k)\displaystyle\sum_{j=l}^{p}\lambda_j\left(z_j^{(l)}(k)\right)^2}\right\}^2 \lambda_i$$

$$= \sum_{i=l}^{p} \left[1 - \xi + \frac{\xi\lambda_i}{\displaystyle\sum_{j=l}^{p}\lambda_j\left(z_j^{(l)}(k)\right)^2}\right]^2 \left(z_i^{(l)}(k)\right)^2 \lambda_i + \Psi_i^{(l)}(k),$$

where

$$\Psi_i^{(l)}(k) = 2\left[1 - \xi + \frac{\xi\lambda_i}{\displaystyle\sum_{j=l}^{p}\lambda_j\left(z_j^{(l)}(k)\right)^2}\right]z_i^{(l)}(k)\Phi_i^l(k)\lambda_i$$

$$+ \left(\Phi_i^l(k)\right)^2 \lambda_i.$$

Meanwhile,

$$\Phi_i^l(k) = \phi_{li}(k) - \frac{\xi \lambda_i z_i^{(l)}(k) \sum_{j=1}^{l-1} \lambda_j \left(z_j^{(l)}(k) \right)^2}{w_l^T(k) C w_l(k) \sum_{j=l}^{p} \lambda_j \left(z_j^{(l)}(k) \right)^2}.$$

Since $\lim_{k \to \infty} \phi_{li} = 0, (i = l, \ldots, p)$, by Lemmas 4.6 and 4.7, we have $\lim_{k \to \infty} \Phi_i^l(k) = 0$. So, it follows that $\lim_{k \to \infty} \Psi_i^{(l)}(k) = 0$. Thus, there must exist a constant $\hat{K}_l$ so that

$$w_l^T(k+1) C w_l(k+1)$$

$$\geq \sum_{i=l}^{p} \left[1 - \xi + \frac{\xi \lambda_i}{\sum_{j=l}^{p} \lambda_j \left(z_j^{(l)}(k) \right)^2} \right]^2 \left(z_i^{(l)}(k) \right)^2 \lambda_i$$

$$\geq \left[1 - \xi + \frac{\xi \lambda_p}{\sum_{j=l}^{p} \lambda_j \left(z_j^{(l)}(k) \right)^2} \right]^2 \sum_{i=l}^{p} \left(z_i^{(l)}(k) \right)^2 \lambda_i$$

$$\geq \min_{s>0} \left[1 - \xi + \frac{\xi \lambda_p}{s} \right]^2 s,$$

for all $k > \hat{K}_l$. By Lemma 4.1, it holds that

$$w_l^T(k+1) C w_l(k+1) \geq 4(1 - \xi) \xi \lambda_p,$$

for $k \geq \hat{K}_l$. Since $\lambda_1 \| w_l(k) \|^2 \geq w_l^T(k+1) C w_l(k+1)$, it follows that

$$\| w_l(k+1) \| \geq \sqrt{\frac{4(1 - \xi) \xi \lambda_p}{\lambda_1}} > 0,$$

for $k \geq \hat{K}_l$, where $0 < \xi < 1$.

Then, from Lemma 4.6, it is easy to see that

$$\| w_1(k+1) \| \leq (1 - \xi)^k \| w_l(0) \| + \xi \frac{M_l}{\lambda_p} \sum_{i=1}^{k} \frac{(1 - \xi)^{i-1}}{\| w_1(k-i) \|},$$

for all $k \geq K_l$, where $\lambda_p < r_l(k) < \lambda_1$. Since $0 < \xi < 1$, it holds that

$$\|w_l(k+1)\|$$

$$\leq (1-\xi)^k \|w_l(0)\| + \xi \frac{M_l}{\lambda_p} \sqrt{\frac{\lambda_1}{4(1-\xi)\xi\lambda_p}} \sum_{i=1}^{k}(1-\xi)^{i-1}$$

$$\leq \frac{M_l}{\lambda_p} \sqrt{\frac{\lambda_1}{4(1-\xi)\xi\lambda_p}},$$

for $k \geq \max\{K_l, \hat{K}_l\}$. This completes the proof.

This lemma gives a smaller ultimate bound when k is sufficiently increased. Obviously, in this case, the condition $w_l(k) \notin V_0$ is met.

4.4 Simulation and Discussion

Three sets of experiments in this section will be provided to illustrate the convergence of GHA with these non-zero-approaching adaptive learning rates.

4.4.1 Example 1

To observe the convergence trajectory clearly, the convergence properties of (4.4) are illustrated first in the two-dimensional space. Give a symmetric positive definitive matrix in R^2 that

$$C = \begin{bmatrix} 5 & 1 \\ 1 & 3 \end{bmatrix}.$$

Its eigenvalues are 5.4142 and 2.5858. The unit eigenvectors associated with the eigenvalues are $\pm[0.9239, 0.3827]^T$ and $\pm[-0.3827, 0.9239]^T$, respectively. Figures 4.1 and 4.2 show the trajectories starting from the points randomly selected in R^2 converge to the eigenvectors. The upper pictures in Figures 4.1 and 4.2 show the trajectories of w_1, and the bottom pictures show the trajectories of w_2. It is clear that all trajectories converge along right directions.

4.4.2 Example 2

We will further illustrate the convergence of (4.4) in a high-dimensional space. Consider the following covariance matrix C generated randomly.

It is easy to get the eigenvalues of matrix C. They are

$$0.0778, 0.0236, 0.0186, 0.0020, 0.0000, 0.0000.$$

In this simulation, we extract four principal directions associated with eigenvalues $0.0778, 0.0236, 0.0186, 0.0020$, respectively. Figure 4.3 shows the norm

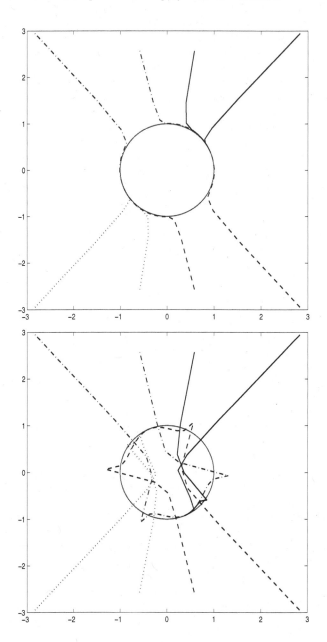

FIGURE 4.1

The trajectories of the first principal direction w_1 converge to the eigenvectors $\pm[0.9239, 0.3827]^T$ (upper) with $\xi = 0.5$, and the trajectories of the second principal direction w_2 converge to the eigenvectors $\pm[-0.3827, 0.9239]^T$ (bottom) with $\xi = 0.5$.

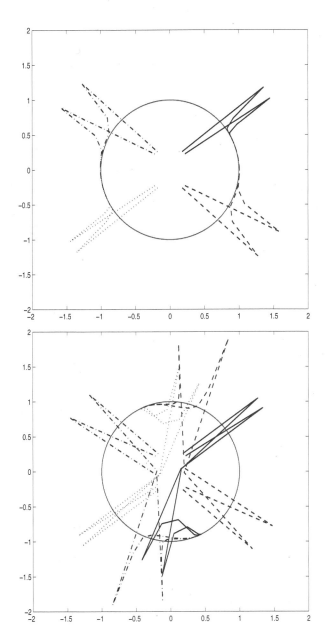

FIGURE 4.2

The trajectories of the first principal direction w_1 converge to the eigenvectors $\pm[0.9239, 0.3827]^T$ (upper) with $\xi = 0.5$, and the trajectories of the second principal direction w_2 converge to the eigenvectors $\pm[-0.3827, 0.9239]^T$ (bottom) with $\xi = 0.5$.

$$
C = \begin{bmatrix}
0.0267 & -0.0030 & -0.0139 & 0.0001 & 0.0317 & 0.0106 \\
-0.0030 & 0.0240 & 0.0016 & -0.0000 & -0.0036 & -0.0012 \\
-0.0139 & 0.0016 & 0.0072 & -0.0001 & -0.0165 & -0.0055 \\
0.0001 & -0.0000 & -0.0001 & 0.0020 & 0.0002 & 0.0001 \\
0.0317 & -0.0036 & -0.0165 & 0.0002 & 0.0377 & 0.0126 \\
0.0106 & -0.0012 & -0.0055 & 0.0001 & 0.0126 & 0.0243
\end{bmatrix}.
$$

evolution of four principal directions with $\xi = 0.3$ or $\xi = 0.7$. Clearly, four principal directions converges quickly to the unit hyper-sphere. Figure 4.4 shows the learning rate evolution of four principal directions with $\xi = 0.3$ or $\xi = 0.7$. From the results, it is shown that learning rates converge to constants. The learning rates make the convergence speed in all eigendirections approximately equal [131] so that the evolution is speeded up considerably with time.

To measure the accuracy of the convergent directions, the direction cosines will be simultaneously computed at the kth update by [32, 31]:

$$
\text{DirectionCosine}_l(k) = \frac{|w_l^T(k) \cdot \mu_l|}{\|w_l(k)\| \cdot \|\mu_l\|},
$$

where μ_l is the true eigenvector associated with the lth eigenvalue of C, which can be computed in advance by MATLAB 7.0. If $DirectionCosine_l(k)$ converges to 1, the principal direction converges to the right direction. Figure 4.5 gives the result.

In above simulations, the initial vectors may be arbitrarily chosen apart from the origin. Simulations show the GHA with this non-zero-approaching learning rates is global convergent, and the convergence speed is quite fast.

4.4.3   Example 3

In this example, we will simulate the convergence of the online version of the algorithm (4.4).

The well-known gray picture for Lenna in Figure 1.5 will be used for the simulation. With the input sequence (1.9), the online observation C_k (1.10) is input to the algorithm.

It is easy to see that $y_l^2(k)$ converges to the λ_l. Thus, on the one hand, the divide by zero condition can be avoided. On the other hand, the convergence speed in all eigendirections is approximately equal as that in [131].

In this simulation, the six principal directions are computed. The convergence results are given in Figures 4.6 and 4.7. It could be seen that the norm of principal directions converge to the unit hyper-sphere in Figure 4.6. Figure 4.7 shows all direction cosines finally approaching 1 with learning rates converging to constants. At the same time, the principal components are extracted. The reconstructed image is shown in Figure 4.8 with SNR of 58.7288.

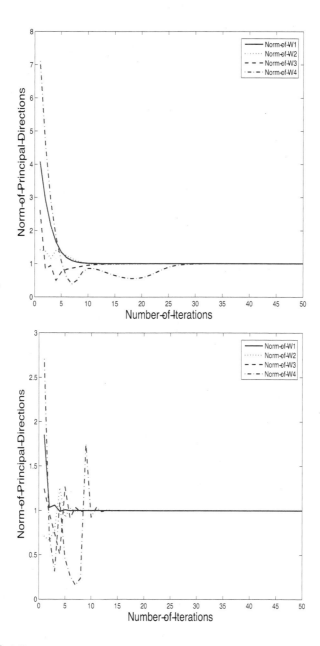

FIGURE 4.3

The norm evolution of four principal directions with $\xi = 0.3$ (upper) or $\xi = 0.7$ (bottom).

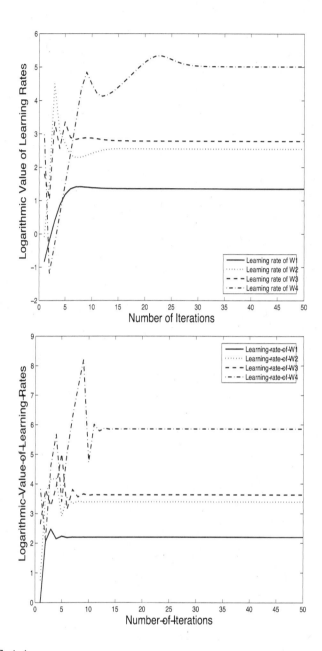

FIGURE 4.4

The learning rate evolution of four principal directions with $\xi = 0.3$ (upper) or $\xi = 0.7$ (bottom).

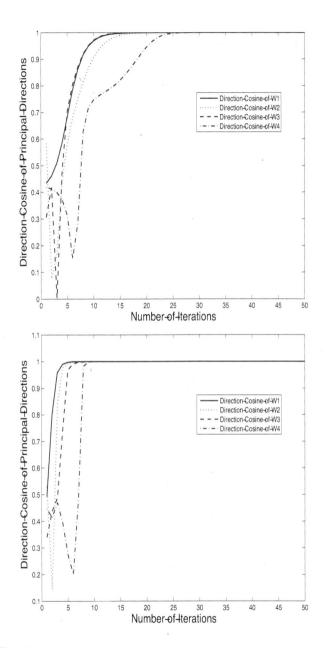

FIGURE 4.5

The direction cosine of four principal directions with $\xi = 0.3$ (upper) or $\xi = 0.7$ (bottom).

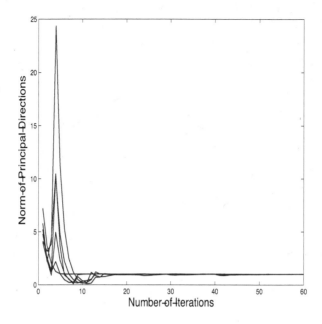

FIGURE 4.6
The norm evolution of six principal directions with $\zeta = 0.5$.

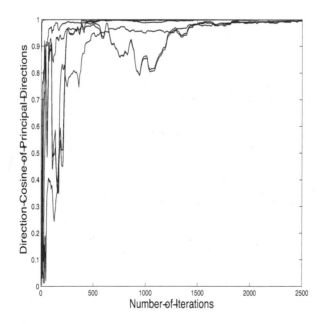

FIGURE 4.7
The direction cosine of six principal directions with $\zeta = 0.5$.

FIGURE 4.8
The reconstructed image (right) with SNR 53.0470.

4.5 Conclusion

This chapter proposes non-zero-approaching adaptive learning rates for the GHA. The convergence of GHA with such learning rates is studied via the DDT method. Since these adaptive learning rates converge to nonzero constants so that, in practical applications, computational round-off limitations and tracking requirements can be satisfied. The learning rates also enable the convergence speed in all eigendirections to be approximately equal so that the evolution does not become slow. In addition, the global convergence of GHA is also guaranteed by using the learning rates. Rigorous mathematical proof is given to prove the global convergence. Simulations are carried out to further illustrate the theory derived.

5

MCA Learning Algorithms

5.1 Introduction

In the past several years, many neural networks learning algorithms were proposed to solve the MCA problem, such as [36, 38, 48, 60, 119, 130, 135, 185]. These learning algorithms can be used to calculate minor component direction from input data without calculating the covariance matrix in advance, which makes neural networks method more suitable in real-time applications and have a lower computational complexity than traditional algebraic approaches, such as eigenvalue decomposition (EVD) and singular value decomposition (SVD). However, there is a divergence problem in some existing MCA algorithms [48, 130]. For example, OJAn algorithm [185], MCA EXIN algorithm [48], and LUO algorithm [119] may suffer from the divergence of weight vector norm, as discussed in [48]. To guarantee convergence, some self-stabilizing MCA learning algorithms are proposed, for instance, MOLLER algorithm [130], CHEN algorithm [36], DOUGLAS algorithm [60], and Ye et al.'s algorithms [188]. In these self-stabilizing algorithms, the weight vector of the neuron can be guaranteed to converge to a normalized minor component direction. In this chapter, following the studies in [36, 60, 130, 186], we propose a stable neural networks learning algorithm for minor component analysis, which has a more satisfactory numerical stability compared with some existing MCA algorithms.

Convergence of MCA learning algorithms is crucial to practical applications [142, 143, 144, 190]. Usually, MCA learning algorithms are described by stochastic discrete time (SDT) systems. It is very difficult to study the convergence of the SDT system directly.

In this chapter, we analyze dynamical behaviors of the proposed MCA algorithm via DDT method and obtain the conditions to guarantee convergence [145].

This chapter is organized as follows. In Section 2, a stable MCA algorithm is proposed. In Section 3, dynamics of the proposed algorithm are analyzed via DDT method. Simulations are carried out to further illustrate the theoretical results in Section 4. Conclusions are given in Section 5.

5.2   A Stable MCA Algorithm

Suppose that the input sequence $\{x(k)|x(k) \in R^n (k = 0, 1, 2, \ldots)\}$ is a zero mean stationary stochastic process and consider a linear neuron with the following input output relation:

$$y(k) = w^T(k)x(k), \quad (k = 0, 1, 2, \ldots),$$

where $y(k)$ is the neuron output and $w(k) \in R^n$ is the weight vector of the neuron. In [184], To solve the problem of principal component analysis (PCA), Xu proposed the following PCA learning algorithm:

$$w(k + 1) = w(k) + \eta \left[\left(2 - w^T(k)w(k) \right) y(k)x(k) - y^2(k)w(k) \right], \quad (5.1)$$

where $\eta > 0$ is the learning rate. In Chapter 3, it has been discussed that Xu's algorithm converges to a principal component direction in which input data have the largest variance. Contrary to principal component direction, input data have the smallest variance in the minor component direction. Thus, a corresponding MCA learning algorithm can be obtained by simply reversing the sign in (5.1) and given as

$$w(k + 1) = w(k) - \eta \left[\left(2 - w^T(k)w(k) \right) y(k)x(k) - y^2(k)w(k) \right]. \quad (5.2)$$

Unfortunately, theoretical analysis and simulation results show that the above algorithm (5.2) is divergent, that is, the norm of weight vector $w(k)$ will approach to infinity. To avoid the problem of divergence, we add a penalty term $0.5 \left[1 - w^T(k)w(k) \right] w(k)$ to (5.2) and get a stable MCA learning algorithm as follows:

$$\begin{aligned} w(k + 1) &= w(k) + 0.5 \left[1 - w^T(k)w(k) \right] w(k) \\ &+ \eta \left(w^T(k)w(k) - 2 \right) y(k)x(k) + \eta y^2(k)w(k). \end{aligned} \quad (5.3)$$

The motivations of adding this penalty term include (1) reducing the deviation of the weight vector norm from unity, that is, guaranteeing that the proposed algorithm is bounded and (2) improving the stability of learning algorithm, which will be illustrated in Theorem 5.3 in Section 3.

By applying the conditional expectation operator $E\{w(k+1)/w(0), x(i), i < k\}$ to (5.3) and identifying the conditional expected value as the next iterate, a DDT system can be obtained and given as

$$\begin{aligned} w(k + 1) &= w(k) + 0.5 \left[1 - w^T(k)w(k) \right] w(k) + \eta \left(w^T(k)w(k) - 2 \right) Cw(k) \\ &+ \eta w^T(k)Cw(k)w(k), \end{aligned} \quad (5.4)$$

where $C = E[x(k)x^T(k)]$ is the covariance matrix of the input sequence $\{x(k)|x(k) \in R^n (k = 0, 1, 2, \ldots)\}$. In this chapter, we analyze the dynamics of (5.4) to interpret the convergence of the algorithm (5.3) indirectly. The

DDT system (5.4) is an "average" of the original stochastic system (5.3). The convergence analysis results about (5.4) could reflect some "average" dynamical behaviors of (5.3).

For convenience of analysis, next, some preliminaries are given. Since the covariance matrix C is a symmetric positive definite matrix, there exists an orthonormal basis of R^n composed of the eigenvectors of C. Let $\lambda_1, \ldots, \lambda_n$ be all the eigenvalues of C ordered by $\lambda_1 > \ldots > \lambda_n > 0$. Suppose that $\{v_i | i = 1, 2, \ldots, n\}$ is an orthonormal basis of R^n such that each v_i is a unit eigenvector of C associated with the eigenvalue λ_i. Thus, for each $k \geq 0$, the weight vector $w(k)$ can be represented as

$$w(k) = \sum_{i=1}^{n} z_i(k) v_i, \tag{5.5}$$

where $z_i(k)(i = 1, 2, \ldots, n)$ are some constants. From (5.4) and (5.5), it holds that for all $k \geq 0$

$$z_i(k+1) = \left[1.5 + (\eta\lambda_i - 0.5) w^T(k)w(k) + \eta w^T(k)Cw(k) - 2\eta\lambda_i\right] z_i(k), \tag{5.6}$$

$(i = 1, 2, \ldots, n)$. According to the relevant properties of the Rayleigh Quotient, it holds that

$$\lambda_n w^T(k)w(k) \leq w^T(k)Cw(k) \leq \lambda_1 w^T(k)w(k), \tag{5.7}$$

for all $k \geq 0$.

5.3 Dynamical Analysis

In this section, we study the dynamical behaviors of DDT system (5.4) via the following three steps:

Step 1: in Theorem 5.1, we prove that if some mild conditions are satisfied, weight vector in (5.4) is bounded.

Step 2: in Theorem 5.2, it is proved that weight vector in (5.4) will converge to minor component under some conditions.

Step 3: in Theorem 5.3, we analyze the stability of DDT system (5.4).

To analyze convergence of DDT system (5.4), we need to prove a preliminary conclusion in the following Lemma 5.1, where it is proved that under some conditions,

$$1.5 + (\eta\lambda_i - 0.5) w^T(k)w(k) + \eta w^T(k)Cw(k) - 2\eta\lambda_i \geq 0,$$

for any $i(i = 1, 2, \ldots, n)$ in (5.6). This preliminary conclusion means that the projection of weight vector $w(k)$ on eigenvector $v_i(i = 1, 2, \ldots, n)$, which is denoted as $z_i(k) = w^T(k)v_i(i = 1, 2, \ldots, n)$, does not change its sign in (5.6).

Lemma 5.1 *Suppose that $\eta\lambda_1 < 0.25$ and $\|w(k)\|^2 \leq 1 + \dfrac{2}{1 - 4\eta\lambda_n}$, then*

$$1.5 + (\eta\lambda_i - 0.5)\, w^T(k)w(k) + \eta w^T(k)Cw(k) - 2\eta\lambda_i \geq 0,$$

($i = 1, \ldots, n$), for all $k \geq 0$.

Proof : Two cases will be considered to complete the proof.

Case 1: $2 \leq \|w(k)\|^2 \leq 1 + \dfrac{2}{1 - 4\eta\lambda_n}$.

By $\eta\lambda_1 \leq 0.25$, it follows from (5.7) that

$$
\begin{aligned}
&\quad\; 1.5 + (\eta\lambda_i - 0.5)\, w^T(k)w(k) + \eta w^T(k)Cw(k) - 2\eta\lambda_i \\
&= 1 + (0.25 - \eta\lambda_i) \cdot \left[2 - w^T(k)w(k)\right] - 0.25 w^T(k)w(k) + \eta w^T(k)Cw(k) \\
&\geq 1 + (0.25 - \eta\lambda_n) \cdot \left[2 - w^T(k)w(k)\right] - 0.25 w^T(k)w(k) \\
&\quad\; + \eta\lambda_n w^T(k)w(k) \\
&= 1.5 - 2\eta\lambda_n - (0.5 - 2\eta\lambda_n) \cdot \|w(k)\|^2 \\
&\geq 1.5 - 2\eta\lambda_n - (0.5 - 2\eta\lambda_n) \cdot \left(1 + \dfrac{2}{1 - 4\eta\lambda_n}\right) \\
&= 0.
\end{aligned}
$$

Case 2: $\|w(k)\|^2 < 2$.

By $\eta\lambda_1 \leq 0.25$, it follows from (5.7) that

$$
\begin{aligned}
&\quad\; 1.5 + (\eta\lambda_i - 0.5)\, w^T(k)w(k) + \eta w^T(k)Cw(k) - 2\eta\lambda_i \\
&= 1 + (0.25 - \eta\lambda_i) \cdot \left[2 - w^T(k)w(k)\right] - 0.25 w^T(k)w(k) + \eta w^T(k)Cw(k) \\
&\geq 1 - 0.25 w^T(k)w(k) + \eta w^T(k)Cw(k) \\
&\geq 1 - (0.25 - \eta\lambda_n)\, w^T(k)w(k) \\
&> 0.5 + 2\eta\lambda_n \\
&> 0.
\end{aligned}
$$

Using the analysis in the previous two cases, we complete the proof.

Next, we prove an interesting lemma that provides an invariant set for DDT system (5.4).

Lemma 5.2 *Denote*

$$S = \left\{ w(k) \,\middle|\, w(k) \in R^n \text{ and } \|w(k)\|^2 \leq 1 + \dfrac{2}{1 - 4\eta\lambda_n} \right\}.$$

If

$$\eta\lambda_1 < 0.25,$$

then S is an invariant set of (5.4).

Proof : To complete the proof, two cases will be considered.

Case 1: $2 \leq \|w(k)\|^2 \leq 1 + \dfrac{2}{1 - 4\eta\lambda_n}$.

By $\eta\lambda_1 < 0.25$, it holds from (5.7) that

$$
\begin{aligned}
&1.5 + (\eta\lambda_i - 0.5)\, w^T(k)w(k) + \eta w^T(k)Cw(k) - 2\eta\lambda_i \\
=\ & 1 + (0.25 - \eta\lambda_i) \cdot \left[2 - w^T(k)w(k)\right] - 0.25 w^T(k)w(k) + \eta w^T(k)Cw(k) \\
\leq\ & 1 - 0.25 w^T(k)w(k) + \eta\lambda_1 w^T(k)w(k) \\
<\ & 1, (i = 1, \ldots, n).
\end{aligned}
\tag{5.8}
$$

Using Lemma 5.1, it follows from (5.6) and (5.8) that

$$
\begin{aligned}
&\|w(k+1)\|^2 \\
=\ & \sum_{i=1}^{n} z_i^2(k+1) \\
=\ & \sum_{i=1}^{n} \left[1.5 + (\eta\lambda_i - 0.5)\, w^T(k)w(k) + \eta w^T(k)Cw(k) - 2\eta\lambda_i\right]^2 \cdot z_i^2(k) \\
\leq\ & \|w(k)\|^2 \\
\leq\ & 1 + \frac{2}{1 - 4\eta\lambda_n}.
\end{aligned}
\tag{5.9}
$$

Case 2: $\|w(k)\|^2 < 2$.

By $\eta\lambda_1 < 0.25$ and Lemma 1, it follows from (5.6) and (5.7) that

$$
\begin{aligned}
&\|w(k+1)\|^2 \\
=\ & \sum_{i=1}^{n} z_i^2(k+1) \\
=\ & \sum_{i=1}^{n} \left[1.5 + (\eta\lambda_i - 0.5)\, w^T(k)w(k) + \eta w^T(k)Cw(k) - 2\eta\lambda_i\right]^2 \cdot z_i^2(k) \\
=\ & \sum_{i=1}^{n} \left\{ 1 + (0.25 - \eta\lambda_i) \cdot \left[2 - w^T(k)w(k)\right] \right. \\
& \left. \qquad -0.25 w^T(k)w(k) + \eta w^T(k)Cw(k) \right\}^2 \cdot z_i^2(k) \\
\leq\ & \left[1 + (0.25 - \eta\lambda_n)\left(2 - \|w(k)\|^2\right) - (0.25 - \eta\lambda_1)\|w(k)\|^2\right]^2 \cdot \|w(k)\|^2 \\
\leq\ & \left[1 + (0.25 - \eta\lambda_n)\left(2 - \|w(k)\|^2\right)\right]^2 \cdot \|w(k)\|^2 \\
\leq\ & \max_{0 \leq s \leq 1 + \frac{2}{1 - 4\eta\lambda_n}} \left[1 + (0.25 - \eta\lambda_n) \cdot (2 - s)\right]^2 \cdot s.
\end{aligned}
\tag{5.10}
$$

By $\eta\lambda_1 < 0.25$, it can be calculated out that for all $s \in \left[0, 1 + \dfrac{2}{1 - 4\eta\lambda_n}\right]$,

$$[1 + (0.25 - \eta\lambda_n) \cdot (2 - s)]^2 \cdot s \leq \frac{16\,(1.5 - 2\eta\lambda_n)^3}{27(1 - 4\eta\lambda_n)}. \tag{5.11}$$

By $\eta\lambda_1 < 0.25$, it holds from (5.10) and (5.11) that

$$\|w(k + 1)\|^2 \leq \frac{16\,(1.5 - 2\eta\lambda_n)^3}{27(1 - 4\eta\lambda_n)} < \frac{2}{1 - 4\eta\lambda_n} < 1 + \frac{2}{1 - 4\eta\lambda_n}. \tag{5.12}$$

From (5.9) and (5.12), we can draw the conclusion that if $\eta\lambda_1 < 0.25$ and $w(k) \in S$, then $\|w(k + 1)\|^2 \leq 1 + \dfrac{2}{1 - 4\eta\lambda_n}$, that is, $w(k + 1) \in S$. The proof is completed.

Next, using Lemma 5.1 and Lemma 5.2, we prove that if some mild conditions are satisfied, the squared norm $\|w(k)\|^2$ of weight vector will be less than a constant $\zeta(0 < \zeta < 2)$ after some iterations in (5.4).

Theorem 5.1 *If $\eta\lambda_1 < 0.25$ and $w(0) \in S$, then there exist a constant $\zeta < 2$ and a positive integer N, such that*

$$\|w(k)\|^2 \leq \zeta,$$

for all $k \geq N$.

Proof: Using Lemma 2, S is an invariant set, then $w(k) \in S$ for all $k \geq 0$. By $\eta\lambda_1 < 0.25$, clearly,

$$0 < \frac{1 - 4\eta\lambda_n}{1 - 2\eta\lambda_1 - 2\eta\lambda_n} < 2.$$

Next, three cases will be considered to complete the proof.

Case 1: $0 \leq \|w(k)\|^2 \leq \dfrac{1 - 4\eta\lambda_n}{1 - 2\eta\lambda_1 - 2\eta\lambda_n}.$

By $\eta\lambda_1 < 0.25$ and Lemma 5.1, it follows from (5.6) and (5.7) that

$$\|w(k+1)\|^2$$

$$= \sum_{i=1}^{n}[1.5 + (\eta\lambda_i - 0.5)\,w^T(k)w(k) + \eta w^T(k)Cw(k) - 2\eta\lambda_i]^2 \cdot z_i^2(k)$$

$$= \sum_{i=1}^{n}\left\{1.5 - 0.5w^T(k)w(k) + \eta w^T(k)Cw(k) - \left[2 - w^T(k)w(k)\right]\eta\lambda_i\right\}^2 \cdot z_i^2(k)$$

$$\leq \sum_{i=1}^{n}\left[1.5 - 0.5\|w(k)\|^2 + \eta\lambda_1\|w(k)\|^2 - \left(2 - \|w(k)\|^2\right)\eta\lambda_n\right]^2 \cdot z_i^2(k)$$

$$= \left[1.5 - 2\eta\lambda_n - (0.5 - \eta\lambda_n - \eta\lambda_1)\|w(k)\|^2\right]^2 \cdot \|w(k)\|^2$$

$$= \left[1 + \frac{1 - 4\eta\lambda_n}{2} \cdot \left(1 - \frac{1 - 2\eta\lambda_1 - 2\eta\lambda_n}{1 - 4\eta\lambda_n} \cdot \|w(k)\|^2\right)\right]^2 \cdot \|w(k)\|^2$$

$$\leq \left[1 + \frac{1}{2} \cdot \left(1 - \frac{1 - 2\eta\lambda_1 - 2\eta\lambda_n}{1 - 4\eta\lambda_n} \cdot \|w(k)\|^2\right)\right]^2 \cdot \|w(k)\|^2$$

$$\leq \left(1.5 - \frac{1 - 2\eta\lambda_n - 2\eta\lambda_1}{2 - 8\eta\lambda_n} \cdot \|w(k)\|^2\right)^2 \cdot \|w(k)\|^2$$

$$\leq \max_{0 \leq s \leq \frac{1 - 4\eta\lambda_n}{1 - 2\eta\lambda_1 - 2\eta\lambda_n}} \left(1.5 - \frac{1 - 2\eta\lambda_n - 2\eta\lambda_1}{2 - 8\eta\lambda_n} \cdot s\right)^2 \cdot s.$$

By $\eta\lambda_1 < 0.25$, it can be calculated out that

$$\left(1.5 - \frac{1 - 2\eta\lambda_n - 2\eta\lambda_1}{2 - 8\eta\lambda_n} \cdot s\right)^2 \cdot s \leq \frac{1 - 4\eta\lambda_n}{1 - 2\eta\lambda_1 - 2\eta\lambda_n},$$

for all $s \in \left[0, \dfrac{1 - 4\eta\lambda_n}{1 - 2\eta\lambda_1 - 2\eta\lambda_n}\right]$. Thus, $\|w(k+1)\|^2 \leq \dfrac{1 - 4\eta\lambda_n}{1 - 2\eta\lambda_1 - 2\eta\lambda_n}$. Thus, it holds that, if

$$\|w(0)\|^2 \leq \frac{1 - 4\eta\lambda_n}{1 - 2\eta\lambda_1 - 2\eta\lambda_n},$$

then

$$\|w(k)\|^2 \leq \frac{1 - 4\eta\lambda_n}{1 - 2\eta\lambda_1 - 2\eta\lambda_n} < 2, (k \geq 0).$$

Case 2: $\dfrac{1 - 4\eta\lambda_n}{1 - 2\eta\lambda_1 - 2\eta\lambda_n} < \|w(k)\|^2 < 2$. By $\eta\lambda_1 < 0.25$, it follows from (5.7) that

$$1.5 + (\eta\lambda_i - 0.5)\,w^T(k)w(k) + \eta w^T(k)Cw(k) - 2\eta\lambda_i$$

$$= 1.5 - 0.5w^T(k)w(k) + \eta w^T(k)Cw(k) - \left[2 - w^T(k)w(k)\right]\eta\lambda_i$$

$$\leq 1.5 - 0.5\|w(k)\|^2 + \eta\lambda_1\|w(k)\|^2 - \left(2 - \|w(k)\|^2\right)\eta\lambda_n$$

$$= 1.5 - 2\eta\lambda_n - (0.5 - \eta\lambda_1 - \eta\lambda_n)\|w(k)\|^2$$

$$\leq 1.$$

Then, using Lemma 5.1, it follows from (5.6) that

$$\|w(k+1)\|^2$$
$$= \sum_{i=1}^{n}[1.5 + (\eta\lambda_i - 0.5)\,w^T(k)w(k) + \eta w^T(k)Cw(k) - 2\eta\lambda_i]^2 \cdot z_i^2(k)$$
$$\leq \|w(k)\|^2.$$

This means that, if $\dfrac{1 - 4\eta\lambda_n}{1 - 2\eta\lambda_1 - 2\eta\lambda_n} < \|w(k)\|^2 < 2$, then $\|w(k)\|^2$ is monotone decreasing. Using the analysis of Case 1, we can obtain that, if

$$\frac{1 - 4\eta\lambda_n}{1 - 2\eta\lambda_1 - 2\eta\lambda_n} < \|w(0)\|^2 < 2,$$

then

$$\|w(k)\|^2 \leq \|w(0)\|^2 < 2, (k \geq 0).$$

Case 3: $2 \leq \|w(k)\|^2 \leq 1 + \dfrac{2}{1 - 4\eta\lambda_n}$.

By $\eta\lambda_1 < 0.25$ and $2 \leq \|w(k)\|^2$, it follows from (5.7) that

$$1.5 + (\eta\lambda_i - 0.5)\,w^T(k)w(k) + \eta w^T(k)Cw(k) - 2\eta\lambda_i$$
$$= 1 + (0.25 - \eta\lambda_i)\cdot(2 - \|w(k)\|^2) - 0.25\|w(k)\|^2 + \eta w^T(k)Cw(k)$$
$$\leq 1 - 0.25\|w(k)\|^2 + \eta\lambda_1\|w(k)\|^2$$
$$= 1 - (0.25 - \eta\lambda_1)\|w(k)\|^2$$
$$\leq 0.5 + 2\eta\lambda_1. \tag{5.13}$$

Using Lemma 5.1, it follows from (5.6) and (5.13) that

$$\|w(k+1)\|^2$$
$$= \sum_{i=1}^{n} z_i^2(k+1)$$
$$= \sum_{i=1}^{n}\left[1.5 + (\eta\lambda_i - 0.5)\,w^T(k)w(k) + \eta w^T(k)Cw(k) - 2\eta\lambda_i\right]^2 \cdot z_i^2(k)$$
$$\leq (0.5 + 2\eta\lambda_1)^2 \cdot \|w(k)\|^2$$
$$= \beta^2 \cdot \|w(k)\|^2$$
$$\leq \beta^{2(k+1)} \cdot \|w(0)\|^2,$$

where $\beta = 0.5 + 2\eta\lambda_1$. Since $\eta\lambda_1 < 0.25$, clearly, $0 < \beta < 1$. Thus, there must exist a positive integer N, such that $\|w(N)\|^2 < 2$. Then, using the analysis of Case 1 and Case 2, we can obtain that for all $k \geq N$,

$$\|w(k)\|^2 \leq \max\left\{\|w(N)\|^2, \frac{1 - 4\eta\lambda_n}{1 - 2\eta\lambda_1 - 2\eta\lambda_n}\right\} < 2.$$

Using the analysis of Case 1, Case 2, and Case 3, we completes the proof.

At this point, the boundedness of DDT system (5.4) has been proved. Next, we prove that under some mild conditions,

$$\lim_{k \to \infty} w(k) = \pm v_n, \tag{5.14}$$

in (5.4), where v_n is minor component direction, that is, the unit eigenvector associated with the smallest eigenvalue of the covariance matrix. From (5.5), for each $k \geq 0$, $w(k)$ can be represented as

$$w(k) = \sum_{i=1}^{n-1} z_i(k)v_i + z_n(k)v_n, \tag{5.15}$$

where $z_i(k)(i = 1, \ldots, n)$ are some constants. Clearly, the convergence of $w(k)$ can be determined by the convergence of $z_i(k)(i = 1, \ldots, n)$. From (5.15), the convergence result (5.14) can be obtained via proving the following two conclusions:

Conclusion 1: $\lim_{k \to \infty} z_i(k) = 0, (i = 1, 2, \ldots, n-1)$,

Conclusion 2: $\lim_{k \to \infty} z_n(k) = \pm 1$.

The following Lemma 5.3 and Lemma 5.4 will provide the proof about *Conclusion 1* and *Conclusion 2*, respectively.

Lemma 5.3 *Suppose that $\eta \lambda_1 < 0.25$, if $w(0) \in S$ and $w^T(0)v_n \neq 0$, then*

$$\lim_{k \to \infty} z_i(k) = 0, (i = 1, 2, \ldots, n-1).$$

Proof : By $\eta \lambda_1 < 0.25$, it holds from (5.7) that

$$\begin{aligned} & \eta w^T(k)Cw(k) - 0.25w^T(k)w(k) \\ \leq\ & \eta \lambda_1 w^T(k)w(k) - 0.25w^T(k)w(k) \\ \leq\ & 0. \end{aligned} \tag{5.16}$$

Using Lemma 5.1, we have that for any $i(i = 1, \ldots, n)$

$$1.5 + (\eta \lambda_i - 0.5)\, w^T(k)w(k) + \eta w^T(k)Cw(k) - 2\eta \lambda_i \geq 0, \tag{5.17}$$

for all $k \geq 0$. Using Theorem 5.1, there must exist a constant $\zeta < 2$ and a positive integer N, such that

$$\|w(k)\|^2 \leq \zeta, \tag{5.18}$$

for all $k \geq N$.

By $\eta \lambda_1 < 0.25$, it follows from (5.7), (5.16), (5.17), and (5.18) that for all

$k \geq N$,

$$
\frac{1.5 + (\eta\lambda_i - 0.5)\, w^T(k)w(k) + \eta w^T(k)Cw(k) - 2\eta\lambda_i}{1.5 + (\eta\lambda_n - 0.5)\, w^T(k)w(k) + \eta w^T(k)Cw(k) - 2\eta\lambda_n}
$$

$$
\leq \frac{1 + (0.25 - \eta\lambda_i) \cdot [2 - w^T(k)w(k)]}{1 + (0.25 - \eta\lambda_n) \cdot [2 - w^T(k)w(k)]}
$$

$$
\leq \frac{1 + (0.25 - \eta\lambda_i) \cdot (2 - \varsigma)}{1 + (0.25 - \eta\lambda_n) \cdot (2 - \varsigma)}
$$

$$
\leq \frac{1 + (0.25 - \eta\lambda_{n-1}) \cdot (2 - \varsigma)}{1 + (0.25 - \eta\lambda_n) \cdot (2 - \varsigma)}
$$

$$
= \gamma, (i = 1, 2, \ldots, n-1), \tag{5.19}
$$

where

$$
\gamma = \frac{1 + (0.25 - \eta\lambda_{n-1}) \cdot (2 - \varsigma)}{1 + (0.25 - \eta\lambda_n) \cdot (2 - \varsigma)}.
$$

Since $\varsigma < 2$ and $\eta\lambda_1 < 0.25$, it holds that

$$
0 < \gamma < 1. \tag{5.20}
$$

By $w^T(0)v_n \neq 0$, clearly, $z_n(0) \neq 0$, and then $z_n(k) \neq 0 (k > 0)$. Using Lemma 5.1, it follows from (5.6) and (5.19) that

$$
\left[\frac{z_i(k+1)}{z_n(k+1)} \right]^2
$$

$$
= \left[\frac{1.5 + (\eta\lambda_i - 0.5)\, w^T(k)w(k) + \eta w^T(k)Cw(k) - 2\eta\lambda_i}{1.5 + (\eta\lambda_n - 0.5)\, w^T(k)w(k) + \eta w^T(k)Cw(k) - 2\eta\lambda_n} \right]^2 \cdot \left[\frac{z_i(k)}{z_n(k)} \right]^2
$$

$$
\leq \gamma^2 \cdot \left[\frac{z_i(k)}{z_n(k)} \right]^2
$$

$$
\leq \gamma^{2(k-N+1)} \cdot \left[\frac{z_i(N)}{z_n(N)} \right]^2, (i = 1, 2, \ldots, n-1),
$$

for all $k \geq N$. From (5.20), clearly,

$$
\lim_{k \to \infty} \frac{z_i(k)}{z_n(k)} = 0, (i = 1, 2, \ldots, n-1).
$$

By the invariance of S, $z_n(k)$ must be bounded, then,

$$
\lim_{k \to \infty} z_i(k) = 0, (i = 1, 2, \ldots, n-1).
$$

The proof is completed.

Lemma 5.4 *Suppose that $\eta\lambda_1 < 0.25$, if $w(0) \in S$ and $w^T(0)v_n \neq 0$, then it holds that*

$$
\lim_{k \to \infty} z_n(k) = \pm 1.
$$

Proof: Using Lemma 5.3, clearly, $w(k)$ will converge to the direction of the minor component direction v_n, as $k \to \infty$. Suppose at time k_0, $w(k)$ has converged to the direction of v_n, that is,

$$w(k_0) = z_n(k_0) \cdot v_n. \tag{5.21}$$

From (5.6), it holds that

$$
\begin{aligned}
z_n(k+1) &= z_n(k) \cdot \left[(1.5 - 2\eta\lambda_n) + (2\eta\lambda_n - 0.5) z_n^2(k) \right] \\
&= z_n(k) \cdot \left[1 + (0.5 - 2\eta\lambda_n)\left(1 - z_n^2(k) \right) \right], \tag{5.22}
\end{aligned}
$$

for all $k \geq k_0$. Since $w(0) \in S$, using Lemma 5.2, it holds that

$$|z_n(k)|^2 \leq \|w(k)\|^2 \leq 1 + \frac{2}{1 - 4\eta\lambda_n},$$

for all $k \geq 0$. By $\eta\lambda_1 < 0.25$, clearly,

$$1 + (0.5 - 2\eta\lambda_n)\left(1 - z_n^2(k) \right) \geq 0, \tag{5.23}$$

for all $k \geq 0$. Thus, from (5.22) and (5.23), it holds that

$$|z_n(k+1)| = |z_n(k)| \cdot \left[1 + (0.5 - 2\eta\lambda_n)\left(1 - z_n^2(k) \right) \right], \tag{5.24}$$

for all $k \geq k_0$. By $\eta\lambda_1 < 0.25$, then

$$\frac{|z_n(k+1)|}{|z_n(k)|} = 1 + (0.5 - 2\eta\lambda_n)\left[1 - z_n^2(k) \right] = \begin{cases} > 1, \text{if } |z_n(k)| < 1 \\ = 1, \text{if } |z_n(k)| = 1 \\ < 1, \text{if } |z_n(k)| > 1 \end{cases}, \tag{5.25}$$

for all $k \geq k_0$. From (5.25), clearly, $|z_n(k)| = 1$ is a potential stable equilibrium point of (5.24). Next, three cases will be considered to complete the proof.

Case 1: $|z_n(k_0)| \leq 1$.

By $\eta\lambda_n \leq \eta\lambda_1 \leq 0.25$ and $|z_n(k_0)| \leq 1$, it can be calculated out from (5.24) that

$$|z_n(k_0 + 1)| = (2\eta\lambda_n - 0.5)|z_n(k_0)|^3 + (1.5 - 2\eta\lambda_n)|z_n(k_0)| \leq 1.$$

Clearly, it holds that $|z_n(k)| \leq 1$ for all $k \geq k_0$. However, it holds from (5.25) that $|z_n(k)|$ is monotone increasing for all $k \geq k_0$. Thus, $|z_n(k)|$ must converge to the equilibrium point 1, as $k \to \infty$.

Case 2: $|z_n(k)| > 1$ for all $k \geq k_0$.

From (5.25), it holds that $|z_n(k)|$ is monotone decreasing for all $k \geq k_0$. However, $|z_n(k)|$ has a low bound 1 for all $k \geq k_0$. Clearly, $|z_n(k)|$ will converge to the equilibrium point 1, as $k \to \infty$.

Case 3: $|z_n(k_0)| > 1$, and there exists a positive integer $M(M > k_0)$, such that $|z_n(M)| \leq 1$.

Since $|z_n(M)| \leq 1$, in the same way as Case 1, it can be proved that $|z_n(k)|$ must converge to the equilibrium point 1, as $k \to \infty$.

From analysis of the previous three cases, we can obtain that

$$\lim_{k \to \infty} |z_n(k)| = 1. \tag{5.26}$$

At the same time, from (5.22) and (5.23), we have that, if $z_n(k_0) \geq 0$, then $z_n(k) \geq 0$ for all $k > k_0$, or if $z_n(k_0) \leq 0$, then $z_n(k) \leq 0$ for all $k > k_0$. Thus, it holds from (5.26) that

$$\lim_{k \to \infty} z_n(k) = \pm 1.$$

This completes the proof.

In Lemma 5.3, we have proved that if $\eta \lambda_1 < 0.25$, $w(0) \in S$, and $w^T(0)v_n \neq 0$, then

$$\lim_{k \to \infty} z_i(k) = 0, (i = 1, 2, \ldots, n - 1). \tag{5.27}$$

In Lemma 5.4, it is proved that if $\eta \lambda_1 < 0.25$, $w(0) \in S$, and $w^T(0)v_n \neq 0$, then

$$\lim_{k \to \infty} z_n(k) = \pm 1. \tag{5.28}$$

Since

$$w(k) = \sum_{i=1}^{n-1} z_i(k)v_i + z_n(k)v_n,$$

from (5.27) and (5.28), we can draw easily the following conclusion:

Theorem 5.2 *if $\eta \lambda_1 < 0.25$, $w(0) \in S$, and $w^T(0)v_n \neq 0$, then in (5.4),*

$$\lim_{k \to \infty} w(k) = \pm v_n.$$

At this point, we have finished the proof about convergence of DDT system (5.4). Next, we further study the stability of (5.4). Clearly, the set of all equilibrium points of (5.4) is

$$\{v_1, \ldots, v_n\} \cup \{-v_1, \ldots, -v_n\} \cup \{0\}.$$

The following Theorem 5.3 will show the stability of the equilibrium points v_n and $-v_n$ in (5.4).

Theorem 5.3 *The equilibrium points v_n and $-v_n$ are locally asymptotically stable if*

$$\eta \lambda_1 < 0.25.$$

Other equilibrium points of (5.4) are unstable.

Proof : Denote

$$\begin{aligned} G(w) &= w(k + 1) \\ &= w(k) + \eta \left[\left(w^T(k)w(k) - 2 \right) Rw(k) + w^T(k)Rw(k)w(k) \right] \\ &+ 0.5 \left[1 - w^T(k)w(k) \right] w(k). \end{aligned}$$

Then, it follows that

$$\frac{\partial G}{\partial w} = I + 2\eta C - \eta w^T(k)w(k)C - \eta w^T(k)Cw(k)I - 4\eta Cw(k)w^T(k), \quad (5.29)$$

where

$$C = \frac{1}{4\eta}I - R,$$

and I is the unity matrix. For the equilibrium point 0, it holds from (5.29) that

$$\left.\frac{\partial G}{\partial w}\right|_0 = I + 2\eta C = 1.5I - 2\eta R = J_0.$$

Denote the ith eigenvalue of J_0 by $\alpha_0^{(i)}$. By $\eta\lambda_1 < 0.25$, it holds that

$$\alpha_0^{(i)} = 1.5 - 2\eta\lambda_i \geq 1.5 - 2\eta\lambda_1 > 1, (i = 1, 2, \ldots, n).$$

According to the Lyapunov theory, the equilibrium point 0 is unstable.

For the equilibrium points $\pm v_j (j = 1, \ldots, n)$, it follows from (5.29) that

$$\left.\frac{\partial G}{\partial w}\right|_{v_j} = (1 + \eta\lambda_j)I - \eta R - (1 - 4\eta\lambda_j)v_jv_j^T = J_j.$$

Denote the ith eigenvalue of J_j by $\alpha_j^{(i)}$. Clearly,

$$\alpha_j^{(i)} = 1 + \eta(\lambda_j - \lambda_i) \text{ if } i \neq j.$$

$$\alpha_j^{(i)} = 4\eta\lambda_j \text{ if } i = j.$$

For any $j \neq n$, it holds that

$$\alpha_j^{(n)} = 1 + \eta(\lambda_j - \lambda_n) > 1.$$

Clearly, the equilibrium points $\pm v_j (j \neq n)$ are unstable. For the equilibrium points $\pm v_n$, since $\eta\lambda_1 < 0.25$, it follows that

$$\alpha_n^{(i)} = 1 + \eta(\lambda_n - \lambda_i) < 1 \text{ if } i \neq n.$$

$$\alpha_n^{(i)} = 4\eta\lambda_n < 4\eta\lambda_1 < 1 \text{ if } i = n.$$

Thus, $\pm v_n$ are asymptotically stable. The proof is completed.

In this section, we have analyzed dynamics of DDT system (5.4) associated with the proposed stable MCA learning algorithm (5.3) and proved that weight vector of (5.4) will converge to stable equilibrium points $\pm v_n$ under some conditions.

5.4   Simulation Results

In this section, we provide two interesting simulation results to illustrate the convergence and stability of the proposed MCA algorithm (5.3) in a deterministic case and a stochastic case, respectively. In these simulations, we compare performance of the proposed MCA algorithm (5.3) with the following MCA algorithms:

(1) MOLLER MCA algorithm [130]

$$w(k + 1) = w(k) - \eta \left[2w^T(k)w(k) - 1 \right] y(k)x(k) + \eta y^2(k)w(k), \qquad (5.30)$$

(2) DOUGLAS MCA algorithm [60]

$$w(k + 1) = w(k) - \eta w^T(k)w(k)w^T(k)w(k)y(k)x(k) + \eta y^2(k)w(k), \qquad (5.31)$$

(3) OJAm MCA algorithm [68]

$$w(k + 1) = w(k) - \eta y(k)x(k) + \frac{\eta y^2(k)w(k)}{w^T(k)w(k)w^T(k)w(k)}. \qquad (5.32)$$

To measure the convergence speed and accuracy of these algorithms, we compute the norm of $w(k)$ and the direction cosine at kth update by [31]:

$$\text{Direction Cosine}(k) = \frac{\left| w^T(k) \cdot v_n \right|}{\|w(k)\| \cdot \|v_n\|},$$

where v_n is the eigenvector associated with the smallest eigenvalue of the covariance matrix R. Clearly, if Direction Cosine(k) converges to 1, then weight vector $w(k)$ must approach the direction of minor component v_n.

First, let us consider performance comparisons in the deterministic case. By taking conditional expectation, the corresponding DDT systems associated with the above three MCA algorithms (5.30)–(5.32) can be obtained as follows:

(1) DDT system associated with MOLLER MCA algorithm (5.30):

$$w(k + 1) = w(k) - \eta \left[2w^T(k)w(k) - 1 \right] Cw(k) + \eta w^T(k)Cw(k)w(k), \qquad (5.33)$$

(2) DDT system associated with DOUGLAS MCA algorithm (5.31):

$$w(k+1) = w(k) - \eta w^T(k)w(k)w^T(k)w(k)Cw(k) + \eta w^T(k)Cw(k)w(k), \qquad (5.34)$$

(3) DDT system associated with OJAm MCA algorithm (5.32):

$$w(k + 1) = w(k) - \eta Cw(k) + \frac{\eta w^T(k)Cw(k)w(k)}{w^T(k)w(k)w^T(k)w(k)}. \qquad (5.35)$$

We randomly generate a 4×4 symmetric positive definite matrix as

$$C^* = \begin{bmatrix} 1.6064 & 0.2345 & 0.6882 & 1.2175 \\ 0.2345 & 1.6153 & 0.3753 & 0.0938 \\ 0.6882 & 0.3753 & 1.8981 & 0.6181 \\ 1.2175 & 0.0938 & 0.6181 & 1.1799 \end{bmatrix}.$$

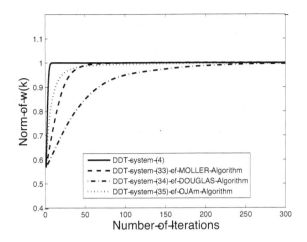

FIGURE 5.1

Convergence of $\|w(k)\|$ in the deterministic case.

We use DDT systems (5.4) and (5.33)–(5.35) to extract the unit eigenvector associated with the smallest eigenvalue of C^* with $\eta = 0.02$ and randomly generated $w(0)$. The simulation results, evaluated over 100 Monte Carlo runs, are presented in Figure 5.1 and Figure 5.2, which show the convergence of $\|w(k)\|$ and Direction Cosine(k), respectively. From the simulation results, we can find that $\|w(k)\|$ in DDT system (5.4) can converge to 1 faster than the other three systems.

Next, let us use stochastic input data to compare stability of the proposed algorithm (5.3) with ones of MOLLER algorithm (5.30), DOUGLAS algorithm (5.31), and OJAm algorithm (5.32). In the simulation, the input data sequence which is generated by [68]

$$x(k) = C \cdot h(k),$$

where $C = \text{randn}(5,5)/5$ and $h(k) \in R^{5 \times 1}$ is Gaussian and randomly produced. The above-mentioned four MCA learning algorithms (5.3), (5.30), (5.31), and (5.32) are used to extract minor component from the input data sequence $\{x(k)\}$. Figure 5.3 and Figure 5.4 show the convergence of $\|w(k)\|$ and Direction Cosine(k), respectively. From the simulation results, we can find easily that in these MCA algorithms, Direction Cosine(k) can converge to 1 at the approximately same speed and weight vector norm $\|w(k)\|$ in the proposed MCA algorithm (5.3) has a better numerical stability than the other three algorithms.

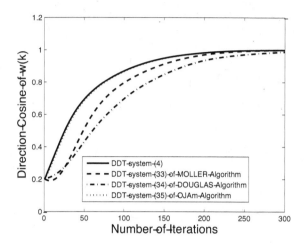

FIGURE 5.2
Convergence of Direction Cosine of $w(k)$ in the deterministic case.

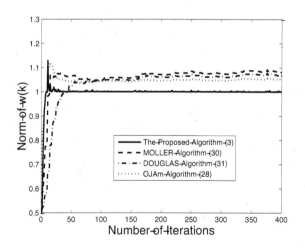

FIGURE 5.3
Convergence of $\|w(k)\|$ in the stochastic case.

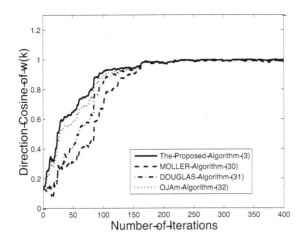

FIGURE 5.4
Convergence of Direction Cosine of $w(k)$ in the stochastic case.

5.5 Conclusion

In this chapter, we propose a stable MCA learning algorithm that has more satisfactory convergence and stability than some existing MCA approaches. Dynamics of the proposed algorithm are analyzed by the deterministic discrete time (DDT) method. It is proved that if some mild conditions about the learning rate and the initial weight vector are satisfied, our proposed algorithm will converge to the minor component direction with unit norm. At the same time, stability analysis shows that minor component direction is the asymptotical stable equilibrium point in the proposed algorithm. Performance comparisons among our proposed algorithm and some existing MCA approaches are carried out in the deterministic case and the stochastic case, which illustrates the performance improvement of the former.

6

ICA Learning Algorithms

6.1 Introduction

Independent component analysis (ICA) can estimate the independent components from the mixed input signal. Its goal is to find a linear representation of nonGaussian data so that the components are statistically independent, or as independent as possible [88].

ICA was originally developed to deal with problems that are closely related to the cocktail-part problem [88]. Since the recent increase of interest in ICA, it has become clear that this principle has a lot of other interesting applications as well. Many learning algorithms were proposed to solve the problem, such as FastICA [85, 88, 138, 149], JADE algorithm [27], EGLD algorithm [65], Pearson-ICA algorithm [98], nonnegative ICA algorithm [147, 187], CuBICA [22, 23].

In this chapter[3], using the DDT method, we study the convergence of a class of Hyvärinen-Oja's ICA learning algorithms with constant learning rates. By directly analyzing the original SDT algorithms, some invariant sets are obtained in stochastic environment so that the nondivergence of the original SDT algorithms is guaranteed by selecting proper learning parameters. In these invariant sets, the local convergence of the original algorithms has been indirectly studied via the DDT method.

To improve the performance of the original SDT algorithms, the corresponding DDT algorithms are extended to the block versions of the original algorithms. The block algorithms not only establish a relationship between the original SDT algorithms and the corresponding DDT algorithms, but also get a good convergence speed and accuracy. The block algorithms represent an average evolution of the original SDT algorithms over all blocks of samples. Thus, the invariant sets of the original SDT algorithms can guarantee the nondivergence of the block algorithms. Furthermore, the block algorithms can exploit more efficiently the information contained in the observed sample window. Therefore, these algorithms converge fast and with higher accuracy [197, 198].

---

[3]Based on "Convergence analysis of Hyvarinen and Oja's ICA learning algorithms with constant learning rates," by (Jian Cheng Lv, K. K. Tan, Zhang Yi, and S. Huang), which appeared in IEEE Trans. Signal Processing, vol. 57, no. 5, pp. 1811–1824, May 2009. ©[2009] IEEE.

This chapter is organized as follows. Section 2 provides the problem formulation and gives preliminaries. Convergence results are given in Section 3. The DDT algorithms are extended to the block algorithms in Section 4. Simulation results and discussions are presented in Section 5. Conclusions are drawn in Section 6.

6.2 Preliminaries and Hyvärinen-Oja's Algorithms

Suppose a sequence of observations $\{\mathbf{y}(k), k = 1, 2, \ldots\}$ with m scalar random variables, that is, $\mathbf{y}(k) = [y_1(k), y_2(k), \ldots, y_m(k)]^T$. In the simplest of ICA [49], it is assumed that

$$\mathbf{y}(k) = A\mathbf{s}(k),$$

where $\mathbf{s}(k) = [s_1(k), s_2(k), \ldots, s_n(k)]^T$ is unknown and $s_1(k), s_2(k), \ldots, s_n(k)$ are mutually and statistically independent with zero mean and unit variance. A is an unknown $m \times n$ mixing matrix of full rank. The basic problem of ICA is then to estimate the original $s_i(k)$ from the mixtures $y(k)$. The problem can be simplified by prewhitening of the data $\mathbf{y}(k)$. The observed data $\mathbf{y}$ is linearly transformed to a vector $\mathbf{x} = M\mathbf{y}$ such that its elements x_i are mutually uncorrelated and all have unit variance, that is, $E\{\mathbf{x}(k)\mathbf{x}^T(k)\} = I$. It follows that

$$\mathbf{x}(k) = M\mathbf{y}(k) = MA\mathbf{s}(k) = B\mathbf{s}(k),$$

where B is an orthogonal matrix. Clearly, $E\{\mathbf{x}(k)\mathbf{x}^T(k)\} = BE\{\mathbf{s}(k)\mathbf{s}^T(k)\}B^T = BB^T = I$, where $B = (\mathbf{b}_1, \mathbf{b}_2, \ldots, \mathbf{b}_n)$, $s_i(k) = \mathbf{b}_i^T x(k)$. Then, the problem of ICA is transformed to one of estimating the $\mathbf{b}_i$.

The kurtosis of the signals is usually used to solve the ICA problem [85, 86, 87]. As for a zero mean random variable v, the kurtosis is defined as $Kurt(v) = E\{v^4\} - 3(E\{v^2\})^2$. It is easy to show that $Kurt(v_1 + v_2) = Kurt(v_1) + Kurt(v_2)$ and $Kurt(\alpha v_1) = \alpha^4 Kurt(v_1)$, where v_1 and v_2 are two independent, zero mean random variables and α is a scalar.

For the signals $s_1(k), s_2(k), \ldots, s_n(k)$, denote by $Kurt(s_i)$ the kurtosis of the signal $s_i(k)$. Clearly, $Kurt(s_i) = E\{s_i^4(k)\} - 3$. Define some sets as follows: $P = \{i|Kurt(s_i) < 0, i \in \{1, 2, \ldots, n\}\}$, $Q = \{i|Kurt(s_i) > 0, i \in \{1, 2, \ldots, n\}\}$, $O = \{i|Kurt(s_i) = 0, i \in \{1, 2, \ldots, n\}\}$ and $T = O \cup P \cup Q$.

Let V_q be the subspace spanned by $\{\mathbf{b}_i|Kurt(s_i) > 0\}$, that is, $V_q = span\{\mathbf{b}_i|Kurt(s_i) > 0\}$. The subspace $V_q^{\perp}$ is perpendicular to V_q. Let $V_p = span\{\mathbf{b}_i|Kurt(s_i) < 0\}$ and $V_p^{\perp}$ is perpendicular to V_p.

To estimate the direction of $\mathbf{b}_i$, in [87], Hyvärinen and Oja proposed the ICA learning algorithms as

$$\mathbf{w}(k+1) = \mathbf{w}(k) + \mu(k) \left[\sigma \mathbf{x}(k) f(\mathbf{w}^T(k)\mathbf{x}(k)) + a(1 - \|\mathbf{w}(k)\|^2)\mathbf{w}(k) \right],$$

for all $k \geq 0$ and $a > 0$, where f is a scalar function, and $\sigma = \pm 1$ is a sign that determines whether we are minimizing or maximizing the kurtosis. $\mu(k)$ is the learning rate sequence and $\mathbf{x}(k)$ is a prewhitened input data. In this chapter, let $f(y) = y^3$, and the constant learning rates are used. The algorithms with constant learning rates can be presented as

$$\mathbf{w}(k+1) = \mathbf{w}(k) + \eta \left[b\mathbf{x}(k)(\mathbf{w}^T(k)\mathbf{x}(k))^3 + a(1 - \|\mathbf{w}(k)\|^2)\mathbf{w}(k) \right], \quad (6.1)$$

$$\mathbf{w}(k+1) = \mathbf{w}(k) + \eta \left[-b\mathbf{x}(k)(\mathbf{w}^T(k)\mathbf{x}(k))^3 + a(1 - \|\mathbf{w}(k)\|^2)\mathbf{w}(k) \right], \quad (6.2)$$

for all $k \geq 0$, where η is a constant learning rate. A constant $b \geq 0$ is added to the algorithms to adjust the learning parameters more conveniently.

The algorithm (6.1) converges to an independent component direction with a positive kurtosis, that is, $\mathbf{w}(k)$ will converge to a direction of a column $\mathbf{b}_i$ of B with $Kurt(s_i) > 0$. The algorithm (6.2) converges to an independent component direction with a negative kurtosis, that is, $\mathbf{w}(k)$ will converge to a direction of a column $\mathbf{b}_i$ of B with $Kurt(s_i) < 0$.

To prove the convergence of the algorithms, a transformation is given first. Let $\mathbf{w}(k) = B\tilde{\mathbf{z}}(k)$. Clearly, to prove that $\mathbf{w}(k)$ converges to a direction of a column $\mathbf{b}_i$ of B, it only needs to prove that all components of $\tilde{\mathbf{z}}(k)$ converge to zero except $\tilde{z}_i(k)$. Since $\mathbf{x}(k) = B\mathbf{s}(k)$, from (6.1) and (6.2), it follows that

$$\tilde{\mathbf{z}}(k+1) = \tilde{\mathbf{z}}(k) + \eta \left[b\mathbf{s}(k)(\tilde{\mathbf{z}}^T(k)\mathbf{s}(k))^3 + a(1 - \|\tilde{\mathbf{z}}(k)\|^2)\tilde{\mathbf{z}}(k) \right], \quad (6.3)$$

$$\tilde{\mathbf{z}}(k+1) = \tilde{\mathbf{z}}(k) + \eta \left[-b\mathbf{s}(k)(\tilde{\mathbf{z}}^T(k)\mathbf{s}(k))^3 + a(1 - \|\tilde{\mathbf{z}}(k)\|^2)\tilde{\mathbf{z}}(k) \right], \quad (6.4)$$

for all $k \geq 0$.

In the following section, the convergence of the Hyvärinen-Oja algorithms will be analyzed in detail. First, some lemmas are provided to be used for subsequent analysis.

Lemma 6.1 *Suppose that $D > 0$, $E > 0$. It holds that*

$$[D - Eh]^2 \cdot h \leq \frac{4D^3}{27E},$$

for all $0 < h \leq \dfrac{D}{E}$.

Proof : Define a differentiable function

$$f(h) = [D - Eh]^2 \cdot h,$$

for $0 < h < \dfrac{D}{E}$. It follows that $\dot{f}(h) = [D - Eh] \cdot [D - 3Eh]$. It is easy to get the maximum point of $f(h)$ on the interval $(0, \frac{D}{E})$ must be $h = \dfrac{D}{3E}$. Thus, it follows that

$$[D - Eh]^2 \cdot h \leq \frac{4D^3}{27E},$$

where $0 < h < \dfrac{D}{E}$. The proof is complete.

Lemma 6.2 *Suppose $H > F > 0$. If $N > C > 0$, then*

$$\frac{F+C}{H+C} < \frac{F+N}{H+N}.$$

Proof: It holds that

$$\frac{F+C}{H+C} - \frac{F+N}{H+N} = \frac{(F-H)(N-C)}{(H+C)(H+N)} < 0.$$

The proof is complete.

Lemma 6.3 *Suppose $0 < \varrho < 1$. If $0 < \eta a < 1$ and $\varrho \le h \le 2$, it holds that*

$$|(1-h)^2 - \eta ah| \le \zeta,$$

where $\zeta = \max\{|(1-\varrho)^2 - \eta a\varrho|, |1 - 2\eta a|\}$ and $0 < \zeta < 1$.

Proof: Define a differentiable function

$$f(h) = (1-h)^2 - \eta ah,$$

for all $\varrho \le h \le 2$, where $0 < \eta a < 1$. It follows that $\dot{f}(h) = -2(1-h) - \eta a$. It is easy to get the maximum point of $f(h)$ on the interval $[\varrho, 2]$ must be $h = \varrho$ or $h = 2$. Thus, it follows that

$$|(1-h)^2 - \eta ah| \le \zeta,$$

where $\zeta = \max\{|(1-\varrho)^2 - \eta a\varrho|, |1 - 2\eta a|\} < 1$. The proof is complete.

6.3   Convergence Analysis

The algorithms (6.1) and (6.2) may diverge. Consider one-dimensional examples. From (6.1), it follows that

$$w(k+1) = w(k) + \eta\left[bw^3(k)x^4(k) + a(1 - w^2(k))w(k)\right]$$
$$= [1 + \eta a - \eta(a - bx^4(k))w(k)^2]w(k),$$

where $\eta > 0, a > 0$ and $b > 0$. Suppose $a - bx^4(k) > \bar{a} > 0$ for all $k \ge 0$, where $\bar{a}$ is a constant. If $w(k) > \dfrac{1 + \eta a}{\eta\bar{a}}$, it holds that $\eta(a - bx^4(k))w(k)^2 - w(k) - \eta a - 1 > 0$. It follows that

$$|w(k+1)| = w^2(k)\frac{\eta(a - bx^4(k))w^2(k) - \eta a - 1}{w(k)}$$
$$> w^2(k),$$

for all $k \geq 0$. Thus, the trajectory approaches toward infinity if $w(0) > \dfrac{1 + \eta a}{\eta \bar{a}}$ under the condition $a - bx^4(k) > \bar{a}$ for all $k \geq 0$. From (6.2), it follows that

$$w(k+1) = w(k) + \eta \left[-bw^3(k)x^4(k) + a(1 - w^2(k))w(k) \right]$$
$$= [1 + \eta a - \eta(a + bx^4(k))w^2(k)]w(k),$$

where $\eta > 0, a > 0$ and $b > 0$. If $w(k) > \dfrac{1 + \eta a}{\eta a}$, it holds that $\eta(a + bx^4(k))w^2(k) - w(k) - \eta a - 1 > 0$. It follows that

$$| w(k+1) | = w^2(k) \frac{\eta(a + bx^4(k))w^2(k) - \eta a - 1}{w(k)}$$
$$> w^2(k),$$

for all $k \geq 0$. Thus, the trajectory approaches toward infinity if $w(0) > \dfrac{1 + \eta a}{\eta a}$.

The examples previous show the algorithms (6.1) and (6.2) may diverge. A problem to address is therefore to find the conditions under which the algorithms are bounded. In the following section, by directly studying the SDT algorithms (6.1) and (6.2), some interesting theorems are given to guarantee the nondivergence of the algorithms in stochastic environment.

6.3.1 Invariant Sets

Let $\Gamma_w = sup \left\{ \| \mathbf{x}(k)\mathbf{x}^T(k) \|^2, k = 1, 2, \ldots \right\}$, assumed Γ_w is finite and $\| \mathbf{w}(0) \|^2 \neq 0$. Suppose $\| \mathbf{w}(0) \|^2 \leq \dfrac{1 + \eta a}{\eta a}$. Define

$$M_{w1} = \min \left\{ \delta_{w1} \left[G - \eta(a - b\Gamma_w) \| \mathbf{w}(0) \|^2 \right]^2 \| \mathbf{w}(0) \|^2, \delta_{w1} \frac{(b\Gamma_w + \eta ab\Gamma_w)^2 G}{\eta a^3} \right\},$$

where $G = 1 + \eta a$, $a > b\Gamma_w$ and $\delta_{w1} > 0$ is a small constant. Clearly, $M_{w1} > 0$. Suppose $\| \mathbf{w}(0) \|^2 \leq \dfrac{1 + \eta a}{\eta a + \eta b\Gamma_w}$. Define

$$M_{w2} = min \left\{ \delta_{w2} \left[G - \eta(a - b\Gamma_w) \| \mathbf{w}(0) \|^2 \right]^2 \| \mathbf{w}(0) \|^2, \delta_{w2} \frac{4(\eta b\Gamma_w)^2 G^3}{(\eta a + \eta b\Gamma_w)^3} \right\},$$

where $G = 1 + \eta a$, $a > b\Gamma_w$ and $\delta_{w2} > 0$ is a small constant. Clearly, $M_{w2} > 0$. Define two invariant sets as

$$\left\{ \begin{array}{l} S_{w1} = \left\{ \mathbf{w} \mid \mathbf{w} \in R^n, M_{w1} \leq \| \mathbf{w} \|^2 \leq \dfrac{1 + \eta a}{\eta a} \right\}, \\[3mm] S_{w2} = \left\{ \mathbf{w} \mid \mathbf{w} \in R^n, M_{w2} \leq \| \mathbf{w} \|^2 \leq \dfrac{1 + \eta a}{\eta a + \eta b\Gamma_w} \right\}. \end{array} \right.$$

Theorem 6.1 *Suppose $\eta > 0$, $a > 0$, and $b > 0$. If $\dfrac{b\Gamma_w}{a} < 0.4074$ and $\eta a < 1$, then S_{w1} is an invariant set of (6.1). If $\dfrac{b\Gamma_w}{a} < 0.2558$ and $\eta a < 1$, then S_{w2} is an invariant set of (6.2).*

Proof : From (6.1), it follows that

$$\mathbf{w}(k+1) = [1+\eta a - \eta a \|\mathbf{w}(k)\|^2]\mathbf{w}(k) + \eta b \mathbf{w}^T(k)\mathbf{x}(k)\mathbf{x}^T(k)\mathbf{w}(k)\mathbf{x}(k)\mathbf{x}^T(k)\mathbf{w}(k),$$

for all $k \geq 0$. Thus, there must exist $\alpha(k)$ so that

$$\|\mathbf{w}(k+1)\| = \alpha(k)\left|1 + \eta a - \eta a \|\mathbf{w}(k)\|^2\right| + \eta b \Gamma_w \|\mathbf{w}(k)\|^2 \big| \|\mathbf{w}(k)\|,$$

for all $k \geq 0$, where $0 < \alpha(k) \leq 1$. Since

$$\alpha(k) = \frac{\|[1 + \eta a - \eta a \|\mathbf{w}(k)\|^2]\mathbf{w}(k) + \eta \mathbf{w}^T(k)\mathbf{x}(k)\mathbf{x}^T(k)\mathbf{w}(k)\mathbf{x}(k)\mathbf{x}^T(k)\mathbf{w}(k)\|}{\|1 + \eta a - \eta a \|\mathbf{w}(k)\|^2\| + \eta b \Gamma_w \|\mathbf{w}(k)\|^2\| \|\mathbf{w}(k)\|},$$

there must exist a small constant δ_{w1} so that $0 < \delta_{w1} < \alpha(k) \leq 1$. So, if $\|\mathbf{w}(k)\|^2 \leq \dfrac{1 + \eta a}{\eta a}$, it holds that

$$\|\mathbf{w}(k+1)\|^2 = \left[\alpha(k)(1 + \eta a) - \alpha(k)(\eta a - \eta b \Gamma_w)\|\mathbf{w}(k)\|^2\right]^2 \|\mathbf{w}(k)\|^2,$$

for all $k \geq 0$, where $0 < \delta_{w1} < \alpha(k) \leq 1$. Since

$$\|\mathbf{w}(k)\|^2 \leq \frac{1 + \eta a}{\eta a} < \frac{\alpha(k)(1 + \eta a) + 1}{\alpha(k)(\eta a - \eta b \Gamma_w)},$$

where $a > b\Gamma_w$, by Lemma 6.1, it follows that

$$\|\mathbf{w}(k+1)\|^2 = \left[\alpha(k)(1 + \eta a) - \alpha(k)(\eta a - \eta b \Gamma_w)\|\mathbf{w}(k)\|^2\right]^2 \|\mathbf{w}(k)\|^2 \quad (6.5)$$
$$\leq \max_{0 < h < \psi(k)} \left\{ [\alpha(k)(1 + \eta a) - \alpha(k)(\eta a - \eta b \Gamma_w)h]^2 h \right\}$$
$$\leq \frac{4\left[\alpha(k)(1 + \eta a)\right]^3}{27\alpha(k)(\eta a - \eta b \Gamma_w)},$$

where $\psi(k) = \dfrac{(1 + \eta a)}{(\eta a - \eta b \Gamma_w)}$. Clearly, if $\dfrac{b\Gamma_w}{a} < 0.4074$ and $\eta a < 1$, it follows that

$$\|\mathbf{w}(k+1)\|^2 \leq \frac{4\left[\alpha(k)(1 + \eta a)\right]^3}{27\alpha(k)(\eta a - \eta b \Gamma_w)} < \frac{1 + \eta a}{\eta a}.$$

From (6.5), it follows that

$$\|\mathbf{w}(k+1)\|^2 \geq \min_{0 < h < \psi(k)} \left\{ [\alpha(k)(1 + \eta a) - \alpha(k)(\eta a - \eta b \Gamma_w)h]^2 h \right\},$$

where $\psi(k) = \dfrac{(1 + \eta a)}{(\eta a - \eta b \Gamma_w)}$. From Lemma 6.1, it is easy to see the minimum

point is $\|\mathbf{w}(k)\|^2 = \|\mathbf{w}(0)\|^2$ or $\|\mathbf{w}(k)\|^2 = \dfrac{1 + \eta a}{\eta a}$ in the interval $\left[0, \dfrac{1 + \eta a}{\eta a}\right]$.

Thus, it can be obtained that $\|\mathbf{w}(k + 1)\|^2 \geq M_{w1} > 0$, for all $k \geq 0$.

From (6.2), there must exist $\beta(k)$ and a small constant δ_{w2} so that

$$\|\mathbf{w}(k + 1)\| = \beta(k) \left| \left| 1 + \eta a - \eta a \|\mathbf{w}(k)\|^2 \right| + \eta b \Gamma_w \|\mathbf{w}(k)\|^2 \right| \|\mathbf{w}(k)\|,$$

for all $k \geq 0$, where $0 < \delta_{w2} < \beta(k) \leq 1$. If $\|\mathbf{w}(k)\|^2 \leq \dfrac{1 + \eta a}{\eta a + \eta b \Gamma_w}$, it holds
that

$$\|\mathbf{w}(k + 1)\|^2 = \left[\beta(k)(1 + \eta a) - \beta(k)(\eta a - \eta b \Gamma_w)\|\mathbf{w}(k)\|^2 \right]^2 \|\mathbf{w}(k)\|^2,$$

for all $k \geq 0$, where $0 < \delta_{w2} < \beta(k) \leq 1$. If $\dfrac{b\Gamma_w}{a} < 0.2558$ and $\eta a < 1$, it
follows that

$$M_{w2} \leq \|\mathbf{w}(k + 1)\|^2 < \dfrac{1 + \eta a}{\eta a + \eta b \Gamma_w},$$

for all $k \geq 0$. The proof is complete.

The previous theorem shows that any trajectory of algorithms (6.1) and (6.2) starting from $\mathbf{w}(0)$ in the invariant sets S_{w1} or S_{w2} will remain in S_{w1} or S_{w2}. This guarantees the non-divergence of the algorithms in stochastic environment.

The invariant sets are obtained by directly studying the original SDT algorithms (6.1) and (6.2). Γ_w is the square of the Frobenius norm and estimated from the available samples. Clearly, the size of samples does not influence the result of Theorem 6.1. However, Theorem 6.1 does not require prewhitening. Thus, the invariant sets of the original algorithms are also suitable for the data, which is not prewhitened.

Since $\|\mathbf{w}(k)\| = \|B\tilde{\mathbf{z}}(k)\| = \|\tilde{\mathbf{z}}(k)\|$, from (6.3) and (6.4), a similar theorem can be obtained as follows. Let $\Gamma_{\tilde{z}} = sup\left\{ \|\mathbf{s}(k)\mathbf{s}^T(k)\|^2, k = 1, 2, \ldots \right\}$,
assumed $\Gamma_{\tilde{z}}$ is finite and $\|\tilde{\mathbf{z}}(0)\|^2 \neq 0$. Suppose $\|\tilde{\mathbf{z}}(0)\|^2 \leq \dfrac{1 + \eta a}{\eta a}$. Define

$$M_{\tilde{z}1} = \min\left\{ \delta_{\tilde{z}1}[G - \eta(a - b\Gamma_{\tilde{z}})\|\tilde{\mathbf{z}}(0)\|^2]^2\|\tilde{\mathbf{z}}(0)\|^2, \delta_{\tilde{z}1}\dfrac{(b\Gamma_{\tilde{z}} + \eta ab\Gamma_{\tilde{z}})^2 G}{\eta a^3} \right\} > 0,$$

where $G = 1 + \eta a$, $\delta_{\tilde{z}1} > 0$ is a small constant. Suppose $\|\tilde{\mathbf{z}}(0)\|^2 \leq \dfrac{1 + \eta a}{\eta a + \eta b \Gamma_{\tilde{z}}}$.
Define

$$M_{\tilde{z}2} = \min\left\{ \delta_{\tilde{z}2}[G - \eta(a - b\Gamma_{\tilde{z}})\|\tilde{\mathbf{z}}(0)\|^2]^2\|\tilde{\mathbf{z}}(0)\|^2, \delta_{\tilde{z}2}\dfrac{4(\eta b\Gamma_{\tilde{z}})^2(1 + \eta a)^3}{(\eta a + \eta b \Gamma_{\tilde{z}})^3} \right\} > 0,$$

where $G = 1 + \eta a$, $\delta_{\tilde{z}2} > 0$ is a small constant. Denote $S_{\tilde{z}1}$ and $S_{\tilde{z}2}$ as two invariant set by

$$\begin{cases} S_{\tilde{z}1} = \left\{ \tilde{z} \mid \tilde{z} \in R^n, M_{\tilde{z}1} \leq \|\tilde{z}\|^2 \leq \dfrac{1 + \eta a}{\eta a} \right\}, \\ S_{\tilde{z}2} = \left\{ \tilde{z} \mid \tilde{z} \in R^n, M_{\tilde{z}2} \leq \|\tilde{z}\|^2 \leq \dfrac{1 + \eta a}{\eta a + \eta b \Gamma_{\tilde{z}}} \right\}. \end{cases}$$

From the analysis of Theorem 6.1, this immediately allows us to state the following theorem.

Theorem 6.2 *Suppose $\eta > 0$, $a > 0$, and $b > 0$. If $\dfrac{b\Gamma_{\tilde{z}}}{a} < 0.4074$ and $\eta a < 1$, then $S_{\tilde{z}1}$ is an invariant set of (6.3). If $\dfrac{b\Gamma_{\tilde{z}}}{a} < 0.2558$ and $\eta a < 1$, then $S_{\tilde{z}2}$ is an invariant set of (6.4).*

Theorems 6.1 and 6.2 give the nondivergent conditions of the algorithms in stochastic environment. It is clear the norm of the algorithms do not go to infinity under these conditions, even in a stochastic environment. In these invariant sets, the local convergence of the algorithms will be further studied in the following subsection.

6.3.2   DDT Algorithms and Local Convergence

The convergence analysis of algorithms (6.1) and (6.2) can be indirectly studied by analyzing the corresponding DDT algorithms in the invariant sets. The DDT algorithms characterize the average evolution of the original algorithms. The convergence of the DDT algorithms can reflect the dynamical behaviors of the original algorithms [195, 210]. Here, we only present the convergence analysis of (6.1). The convergence results of (6.2) can be obtained by using similar method. This chapter also gives the convergence results of (6.2), but the analysis process of its convergence is omitted.

Following Zufiria's method [210], the corresponding DDT algorithms can be obtained by taking the conditional expectation $E\{\mathbf{w}(k+1)/\mathbf{w}(0), \mathbf{w}(i), i < k\}$ on both sides of the algorithm (6.1). Since $E\{\mathbf{w}(k)\} = BE\{\tilde{\mathbf{z}}(k)\}$, the convergence analysis of the DDT algorithm of (6.1) can be transformed to studying the convergence of the DDT algorithm of (6.3). Taking the conditional expectation $E\{\tilde{\mathbf{z}}(k+1)/\tilde{\mathbf{z}}(0), \tilde{\mathbf{z}}(i), i < k\}$ on both sides of (6.3), the corresponding DDT algorithm is written as

$$\mathbf{z}(k+1) = \mathbf{z}(k) + \eta \left[bE\{\mathbf{s}(k)(\tilde{\mathbf{z}}^T(k)\mathbf{s}(k))^3\} + a(1 - \|\mathbf{z}(k)\|^2)\mathbf{z}(k) \right], \quad (6.6)$$

for all $k \geq 0$, where $\mathbf{z}(k) = E\{\tilde{\mathbf{z}}(k)\}$. The DDT algorithm (6.6) preserves the discrete time form of the original algorithm (6.3) and allows a more realistic behavior of the learning gain [210]. Since the DDT algorithm (6.6) represents an average evolution of the algorithm (6.3), the invariant set $S_{\tilde{z}1}$ is also an invariant set of (6.6).

The following provides the convergence of the algorithm (6.6) in detail.

Lemma 6.4 *(Hyvärinen-Oja [85]) For any $i \in T$, it holds that*

$$E\{s_i(\sum_{j}^{n} \tilde{z}_j(k)s_j)^3\} = Kurt(s_i)z_i^3(k) + 3\|\mathbf{z}(k)\|^2 z_i(k),$$

for all $k \geq 0$.

Proof : The following conclusion can be found in [85]. Here the proof is provided. First, it holds that

$$\left(\sum_{j=1}^{n} \tilde{z}_j(k)s_j(k)\right)^3 = \sum_{i=1}^{n} \tilde{z}_i^3(k)s_i^3(k) + \sum_{i=1}^{n}\sum_{j\neq i}^{n} \tilde{z}_i^2(k)s_i^2(k)\tilde{z}_j(k)s_j(k)$$

$$+ 2\sum_{q=1}^{n} \tilde{z}_q(k)s_q(k) \sum_{i\neq j} \tilde{z}_i(k)\tilde{z}_j(k)s_i(k)s_j(k),$$

for all $k \geq 0$. Then, it follows that

$$E\left\{ s_i(k)\left(\sum_{j=1}^{n} \tilde{z}_j(k)s_j(k)\right)^3 \right\}$$

$$= E\left\{ s_i(k)\sum_{i=1}^{n} \tilde{z}_i^3(k)s_i^3(k) + s_i(k)\sum_{i=1}^{n}\sum_{j\neq i}^{n} \tilde{z}_i^2(k)s_i^2(k)\tilde{z}_j(k)s_j(k) \right.$$

$$\left. + 2s_i(k)\sum_{q=1}^{n} \tilde{z}_q(k)s_q(k) \sum_{i\neq j} \tilde{z}_i(k)\tilde{z}_j(k)s_i(k)s_j(k) \right\}$$

$$= E\left\{ s_i^4(k) \right\} z_i^3(k) + E\left\{ s_i(k)\sum_{i=1}^{n}\sum_{j\neq i}^{n} \tilde{z}_i^2(k)s_i^2(k)\tilde{z}_j(k)s_j(k) \right\}$$

$$+ 2E\left\{ s_i(k)\sum_{q=1}^{n} \tilde{z}_q(k)s_q(k) \sum_{i\neq j} \tilde{z}_i(k)\tilde{z}_j(k)s_i(k)s_j(k) \right\}.$$

Since $s_i(i = 1,\ldots,n)$ are mutually independent, it holds that $E\{s_i^2 s_j^2\} = 1$ and $E\{s_i^3 s_j\} = E\{S_i^2 s_j s_l\} = E\{s_i^2 s_j s_l s_m\}$ for four different indices i, j, l, m [85]. It follows that

$$E\left\{ s_i(k)\left(\sum_{j=1}^{n} \tilde{z}_j(k)s_j(k)\right)^3 \right\} = E\left\{ s_i^4(k) \right\} z_i^3(k) + \sum_{j\neq i} z_j^2(k)z_i^2(k) + 2\sum_{j\neq i} z_j^2(k)z_i(k),$$

for all $k \geq 0$. Since $Kurt(s_i) = E\{s_i^4(k)\} - 3$, it follows that

$$E\left\{s_i(k)\left(\sum_{j=1}^{n}\tilde{z}_j(k)s_j(k)\right)^3\right\} = Kurt(s_i)z_i^3(k) + 3z_i^3(k) + 3\sum_{j\neq i}z_j^2(k)z_i(k)$$

$$= Kurt(s_i)z_i^3(k) + 3z_i^3(k) + 3\sum_{j\neq i}z_j^2(k)z_i(k)$$

$$= Kurt(s_i)z_i^3(k) + 3\|\mathbf{z}(k)\|^2 z_i(k),$$

for all $k \geq 0$. The proof is complete.

By the previous lemma, for any component $z_i(k+1)$ of $\mathbf{z}(k+1)$, it holds that

$$
\begin{aligned}
z_i(k+1) &= z_i(k) + \eta\left[bKurt(s_i)z_i(k)^3 + 3b\|\mathbf{z}(k)\|^2 z_i(k) + a(1 - \|\mathbf{z}(k)\|^2)z_i(k)\right]\\
&= \left[1 + \eta bKurt(s_i)z_i(k)^2 + 3\eta b\|\mathbf{z}(k)\|^2 + \eta a(1 - \|\mathbf{z}(k)\|^2)\right]z_i(k)\\
&= \left[1 + \eta a - \eta a\|\mathbf{z}(k)\|^2 + \eta b(Kurt(s_i)z_i(k)^2 + 3\|\mathbf{z}(k)\|^2)\right]z_i(k), \quad (6.7)
\end{aligned}
$$

for all $k \geq 0$.

Lemma 6.5 *Suppose that $\mathbf{z}(0) \in S_{\tilde{z}1}$ and $\Gamma_{\tilde{z}} > \max\{Kurt(s_i) + 3, 3\}$. For any $i \in T$, it holds that*

$$1 + \eta a - \eta a\|\mathbf{z}(k)\|^2 + \eta b(Kurt(s_i)z_i(k)^2 + 3\|\mathbf{z}(k)\|^2) > 0$$

for all $k \geq 0$.

Proof : It follows that

$$1 + \eta a - \eta a\|\mathbf{z}(k)\|^2 + \eta b(Kurt(s_i(k))z_i(k)^2 + 3\|\mathbf{z}(k)\|^2)$$
$$= 1 + \eta a - \eta\left[a - b\left(Kurt(s_i(k))\frac{z_i(k)^2}{\|\mathbf{z}(k)\|^2} + 3\right)\right]\|\mathbf{z}(k)\|^2,$$

for all $k \geq 0$. Since $\Gamma_{\tilde{z}} > \max\{Kurt(s_i) + 3, 3\}$ and $\mathbf{z}(0) \in S_{\tilde{z}1}$, by Theorem 6.2, it follows that

$$1 + \eta a - \eta a\|\mathbf{z}(k)\|^2 + \eta b(Kurt(s_i(k))z_i(k)^2 + 3\|\mathbf{z}(k)\|^2) > 0,$$

for all $k \geq 0$. The proof is complete.

Lemma 6.6 *Suppose that $\mathbf{z}(0) \in S_{\tilde{z}1}$, $\mathbf{z}(0) \notin V_q^{\perp}$ and $\Gamma_{\tilde{z}} > \max\{Kurt(s_i) + 3, 3\}$. For any $z_i(k) \neq 0, i \in P$, if $z_i^2(k) > \bar{\rho} > 0$ for all $k \geq 0$, it holds that*

$$|z_i(k+1)|^2 \leq |z_j(k+1)|^2 \cdot \left[\frac{z_i(0)}{z_j(0)}\right]^2 \cdot e^{-\theta_1 k} < \Pi_1 \cdot e^{-\theta_1 k},$$

where $j \in Q$ and

$$\Pi_1 = \frac{1 + \eta a}{\eta a} \cdot \left[\frac{z_i(0)}{z_j(0)}\right]^2,$$

$$\theta_1 = ln\left(\frac{(1+\eta a)(a+3b)}{(1+\eta a)(a+3b)+\eta ab\bar{\rho}Kurt(s_i)}\right)^2 > 0.$$

Proof: Since $\mathbf{z}(0) \notin V_q^\perp$, there must exist $j \in Q$ so that $z_j(0) \neq 0$. For any $z_i(k) \neq 0, i \in P$, and $z_i^2(k) > \bar{\rho} > 0$ for all $k \geq 0$, from (6.7), it follows that

$$\left[\frac{z_i(k+1)}{z_j(k+1)}\right]^2 = \left[\frac{1+\eta a-\eta a\|\mathbf{z}(k)\|^2+\eta b(Kurt(s_i(k))z_i(k)^2+3\|\mathbf{z}(k)\|^2)}{1+\eta a-\eta a\|\mathbf{z}(k)\|^2+\eta b(Kurt(s_j(k))z_j(k)^2+3\|\mathbf{z}(k)\|^2)}\right]^2 \cdot \left[\frac{z_i(k)}{z_j(k)}\right]^2,$$

for all $k \geq 0$. By Lemma 6.5 and $i \in P$, that is, $Kurt(s_i) < 0$, it follows that

$$\left[\frac{z_i(k+1)}{z_j(k+1)}\right]^2 \leq \left[\frac{1+\eta a-\eta a\|\mathbf{z}(k)\|^2+\eta b(Kurt(s_i)\bar{\rho}+3\|\mathbf{z}(k)\|^2)}{1+\eta a-\eta a\|\mathbf{z}(k)\|^2+3\eta b\|\mathbf{z}(k)\|^2}\right]^2 \cdot \left[\frac{z_i(k)}{z_j(k)}\right]^2$$

$$\leq \left[\frac{1+\eta a+\eta b\bar{\rho}Kurt(s_i)+3\eta b\|\mathbf{z}(k)\|^2}{1+\eta a+3\eta b\|\mathbf{z}(k)\|^2}\right]^2 \cdot \left[\frac{z_i(k)}{z_j(k)}\right]^2,$$

for all $k \geq 0$. Since $\|\mathbf{z}(k)\|^2 < \dfrac{1+\eta a}{\eta a}$, by Lemma 6.2, it follows that

$$\left[\frac{z_i(k+1)}{z_j(k+1)}\right]^2 \leq \left[\frac{1+\eta a+\eta b\bar{\rho}Kurt(s_i)+3\eta b\cdot(1+\eta a)/\eta a}{1+\eta a+3\eta b\cdot(1+\eta a)/\eta a}\right]^2 \cdot \left[\frac{z_i(k)}{z_j(k)}\right]^2$$

$$\leq \left[\frac{(1+\eta a)(a+3b)+\eta ab\bar{\rho}Kurt(s_i)}{(1+\eta a)(a+3b)}\right]^2 \cdot \left[\frac{z_i(k)}{z_j(k)}\right]^2$$

$$\leq \left[\frac{(1+\eta a)(a+3b)+\eta ab\bar{\rho}Kurt(s_i)}{(1+\eta a)(a+3b)}\right]^{2(k+1)} \cdot \left[\frac{z_i(0)}{z_j(0)}\right]^2$$

$$= \left[\frac{z_i(0)}{z_j(0)}\right]^2 \cdot e^{-\theta_1(k+1)},$$

for all $k \geq 0$, where

$$\theta_1 = ln\left(\frac{(1+\eta a)(a+3b)}{(1+\eta a)(a+3b)+\eta ab\bar{\rho}Kurt(s_i)}\right)^2 > 0.$$

Since $\mathbf{z}(0) \in S_{\bar{z}1}$, it holds that

$$|z_i(k+1)|^2 \leq |z_j(k+1)|^2 \cdot \left[\frac{z_i(0)}{z_j(0)}\right]^2 \cdot e^{-\theta_1 k} < \Pi_1 \cdot e^{-\theta_1 k},$$

where

$$\Pi_1 = \frac{1+\eta a}{\eta a} \cdot \left[\frac{z_i(0)}{z_j(0)}\right]^2.$$

The proof is complete.

Lemma 6.7 *Suppose that* $Kurt(s_{j_0}) > 0, (j_0 \in Q)$ *and* $Kurt(s_j) \geq 0, j \in (Q \cup O)/j_0$. *If* $Kurt(s_{j_0})z_{j_0}^2(0) > Kurt(s_j)z_j^2(0) \geq 0$, *there must exist a constant* $\xi > 0$ *so that* $0 < \psi(k) < \xi < 1$ *for all* $k \geq 0$, *where*

$$\psi(k) = \left[\frac{(1 + \eta a)(a + 3b) + \eta ab Kurt(s_j)z_j^2(k)}{(1 + \eta a)(a + 3b) + \eta ab Kurt(s_{j_0})z_{j_0}^2(k)} \right]^2.$$

Proof : For $j_0 \in Q$ and any $j \in (Q \cup O)/j_0$, from (6.7), it follows that

$$\frac{z_j^2(k+1)}{z_{j_0}^2(k+1)} = \gamma(k) \cdot \frac{z_j^2(k)}{z_{j_0}^2(k)}, \tag{6.8}$$

for all $k \geq 0$, where

$$\gamma(k) = \left[\frac{1 + \eta a - \eta a \|\mathbf{z}(k)\|^2 + \eta b(Kurt(s_j)z_j^2(k) + 3\|\mathbf{z}(k)\|^2)}{1 + \eta a - \eta a \|\mathbf{z}(k)\|^2 + \eta b(Kurt(s_{j_0})z_{j_0}^2(k) + 3\|\mathbf{z}(k)\|^2)} \right]^2.$$

If $Kurt(s_{j_0})z_{j_0}^2(k) > Kurt(s_j)z_j^2(k) \geq 0$, by Lemma 6.5, it holds that $0 < \gamma(k) < 1$. From (6.8), it holds that

$$z_{j_0}^2(k+1)z_j^2(k) > z_j^2(k+1)z_{j_0}^2(k) \geq 0.$$

So, it is easy to see that

$$Kurt(s_{j_0})z_{j_0}^2(k+1) > Kurt(s_j)z_j^2(k+1) \geq 0.$$

Thus, if $Kurt(s_{j_0})z_{j_0}^2(0) > Kurt(s_j)z_j^2(0) \geq 0$, it holds that

$$Kurt(s_{j_0})z_{j_0}^2(k+1) > Kurt(s_j)z_j^2(k+1), \tag{6.9}$$

for all $k \geq 0$.

Now, we will prove that if $Kurt(s_{j_0})z_{j_0}^2(0) > Kurt(s_j)z_j^2(0) \geq 0$, there must exist a small constant $\epsilon > 0$ so that $z_{j_0}^2(k) > \epsilon$ for all $k \geq 0$. Assume that this is not true. Then, it holds that $\lim_{k \to +\infty} z_{j_0}^2(k) = 0, j_0 \in Q$. From (6.9), it follows that $\lim_{k \to +\infty} z_j^2(k) = 0, j \in (Q \cup O)/j_0$. By Lemma 6.6, it holds that $\lim_{k \to +\infty} \|\mathbf{z}(k)\|^2 = 0$. This contradicts with Theorem 6.2.

Since $Kurt(s_{j_0})z_{j_0}^2(0) > Kurt(s_j)z_j^2(0) \geq 0$ and $0 < \gamma(k) < 1$ for all $k \geq 0$. From (6.8), it follows that

$$Kurt(s_{j_0})z_{j_0}^2(k+1) - Kurt(s_j)z_j^2(k+1)$$

$$= Kurt(s_{j_0})z_{j_0}^2(k+1)\left(1 - \gamma(k)\frac{Kurt(s_j)z_j^2(k)}{Kurt(s_{j_0})z_{j_0}^2(k)}\right)$$

$$= Kurt(s_{j_0})z_{j_0}^2(k+1)\left(1 - \prod_{l=1}^{k}\gamma(k) \cdot \frac{Kurt(s_j)z_j^2(0)}{Kurt(s_{j_0})z_{j_0}^2(0)}\right)$$

$$> Kurt(s_{j_0})z_{j_0}^2(k+1)\left(1 - \frac{Kurt(s_j)z_j^2(0)}{Kurt(s_{j_0})z_{j_0}^2(0)}\right),$$

for all $k \geq 0$. It follows that

$$\eta ab Kurt(s_{j_0}) z_{j_0}^2(k+1) - \eta ab Kurt(s_j) z_j^2(k+1) > \bar{\epsilon} > 0,$$

where

$$\bar{\epsilon} = \eta ab Kurt(s_{j_0}) \epsilon \left(1 - \frac{Kurt(s_j) z_j^2(0)}{Kurt(s_{j_0}) z_{j_0}^2(0)} \right).$$

It is easy to show that

$$\psi(k) = \frac{(1 + \eta a)(a + 3b) + \eta ab Kurt(s_j) z_j^2(k)}{(1 + \eta a)(a + 3b) + \eta ab Kurt(s_{j_0}) z_{j_0}^2(k)} < \xi,$$

for all $k \geq 0$, where

$$\xi = 1 - \frac{a \bar{\epsilon}}{(1 + \eta a)[a(a + 3b) + \eta ab Kurt(s_{j_0})]} < 1.$$

The proof is complete.

Lemma 6.8 *Suppose that* $\mathbf{z}(0) \in S_{\tilde{z}1}$, $\mathbf{z}(0) \notin V_q^\perp$ *and* $\Gamma_{\tilde{z}} > \max\{Kurt(s_i) + 3, 3\}$. *If* $Kurt(s_{j_0}) z_{j_0}^2(0) > Kurt(s_j) z_j^2(0) \geq 0, j_0 \in Q$ *and* $j \in (Q \cup O)/j_0$, *it holds that*

$$|z_j(k+1)|^2 \leq |z_{j_0}(k+1)|^2 \cdot \left[\frac{z_j(0)}{z_{j_0}(0)} \right]^2 \cdot e^{-\theta_2 k} < \Pi_2 \cdot e^{-\theta_2 k},$$

for all $j \in (Q \cup O)/j_0$, *where*

$$\Pi_2 = \frac{1 + \eta a}{\eta a} \cdot \left[\frac{z_j(0)}{z_{j_0}(0)} \right]^2 \quad and \quad \theta_2 = ln \left[\frac{1}{\xi} \right]^2 > 0.$$

Proof: Since $\mathbf{z}(0) \notin V_q^\perp$, there must exist $j_0(j_0 \in Q)$ so that $z_j(0) \neq 0$ and $Kurt(s_{j_0}) z_{j_0}^2(0) > Kurt(s_j) z_j^2(0), j \in Q/j_0$. From (6.7), it follows that

$$\frac{z_j^2(k+1)}{z_{j_0}^2(k+1)} = \left[\frac{1 + \eta a - \eta a \|\mathbf{z}(k)\|^2 + \eta b(Kurt(s_j(k)) z_j(k)^2 + 3\|\mathbf{z}(k)\|^2)}{1 + \eta a - \eta a \|\mathbf{z}(k)\|^2 + \eta b(Kurt(s_{j_0}(k)) z_{j_0}(k)^2 + 3\|\mathbf{z}(k)\|^2)} \right]^2 \cdot \frac{z_j^2(k)}{z_{j_0}^2(k)},$$

By Lemmas 6.2 and 6.7, it holds that

$$\frac{z_j^2(k+1)}{z_{j_0}^2(k+1)} \leq \xi \cdot \frac{z_j^2(k)}{z_{j_0}^2(k)},$$

for all $k \geq 0$. Thus, it follows that

$$|z_j(k+1)|^2 \leq |z_{j_0}(k+1)|^2 \cdot \left[\frac{z_j(0)}{z_{j_0}(0)} \right]^2 \cdot e^{-\bar{\theta}_1 k} < \bar{\Pi}_1 \cdot e^{-\bar{\theta}_1 k},$$

for all $j \in (Q \cup O)/j_0$, where

$$\Pi_2 = \frac{1 + \eta a}{\eta a} \cdot \left[\frac{z_j(0)}{z_{j_0}(0)} \right]^2 \quad and \quad \theta_2 = ln \left[\frac{1}{\xi} \right]^2 > 0.$$

The proof is complete.

Lemma 6.9 *Suppose that* $\mathbf{z}(0) \in S_{\tilde{z}1}$, $\mathbf{z}(0) \notin V_q^{\perp}$ *and* $\Gamma_{\tilde{z}} > \max\{Kurt(s_i) + 3, 3\}$. *If* $Kurt(s_{j_0})z_{j_0}^2(0) > Kurt(s_i)z_i^2(0), j_0 \in Q$ *and* $i \in T/j_0$, *it holds that*

$$\eta a - V(k+1) \leq \zeta^k \cdot |\eta a - V(0)|,$$

where $V(k) = \eta[a - b(Kurt(s_{j_0}) + 3)]z_{j_0}^2(k)$ *and* $\zeta = max\{|(1-\gamma)^2 - \eta a \gamma, |1 - 2\eta a|\}$, *when* k *is large enough.*

Proof : By Lemmas 6.6 and 6.8, it follows that

$$\lim_{k \to +\infty} z_i(k) = 0, (i \in P \text{ and } Kurt(s_i) < 0),$$

$$\lim_{k \to +\infty} z_j(k) = 0, (j \in (Q \cup O)/j_0 \text{ and } Kurt(s_j) \geq 0),$$

where $j_0 \in Q, Kurt(s_{j_0})z_{j_0}^2(0) > Kurt(s_j)z_j^2(0)$. From (6.7), when k is large enough, it follows that

$$z_{j_0}(k+1) = z_{j_0}(k) + \left[\eta a - \eta a z_{j_0}^2(k) + \eta b(Kurt(s_{j_0})z_{j_0}^2(k) + 3z_{j_0}^2(k))\right]z_{j_0}(k)$$
$$= \{1 + \eta a - \eta[a - b(Kurt(s_{j_0}) + 3)]z_{j_0}^2(k)\}z_{j_0}(k).$$

Since $\mathbf{z}(0) \in S_{\tilde{z}1}$, there must exist a small constant ϱ so that $0 < \varrho < 1$ and $\varrho \leq V(k)$. Thus, it can be checked that

$$0 < \varrho < V(k) < \eta[a - b(Kurt(s_{j_0}) + 3)]\frac{1 + \eta a}{\eta a}$$
$$= \frac{a - b(Kurt(s_{j_0}) + 3)}{a}(1 + \eta a)$$
$$\leq 2,$$

where $0 < \eta a < 1$ and $a > b(Kurt(s_{j_0}) + 3)$. Then, it follows that

$$\eta a - V(k+1) = \eta a - \eta[a - b(Kurt(s_{j_0}) + 3)]z_{j_0}^2(k+1)$$
$$= \eta a - \eta[a - b(Kurt(s_{j_0}) + 3)] \cdot$$
$$\{1 + \eta a - \eta[a - b(Kurt(s_{j_0}) + 3)]z_{j_0}^2(k)\}^2 z_{j_0}^2(k).$$

It is easy to get that

$$\eta a - V(k+1) = [\eta a - V(k)]\{[1 - V(k)]^2 - \eta a V(k)\}$$
$$\leq |\eta a - V(k)||[1 - V(k)]^2 - \eta a V(k)|.$$

By Lemma 6.3, it follows that

$$|\eta a - V(k+1)| \leq |\eta a - V(k)| \cdot |[1 - V(k)]^2 - \eta a V(k)|$$
$$\leq |\eta a - V(k)| \cdot \zeta$$
$$= \zeta^k |\eta a - V(0)|,$$

where $0 < \zeta < 1$. The proof is complete.

Theorem 6.3 *Suppose that* $\mathbf{z}(0) \in S_{\bar{z}1}$, $\mathbf{z}(0) \notin V_q^{\perp}$ *and* $\Gamma_{\bar{z}} > \max\{Kurt(s_i) + 3, 3\}$. *If* $Kurt(s_{j_0})z_{j_0}^2(0) > Kurt(s_i)z_i^2(0), j_0 \in Q$ *and* $i \in T/j_0$, *there must exist a constant* $z_{j_0}^*$ *so that* $\lim\limits_{k \to +\infty} E\{\mathbf{w}(k)\} = B\mathbf{z}(k) = \mathbf{b}_{j_0} \cdot z_{j_0}^*$.

Proof. By Lemmas 6.6 and 6.8, when k is large enough, it follows that

$$z_{j_0}(k+1) = z_{j_0}(k) + \left[\eta a - \eta a z_{j_0}^2(k) + \eta b(Kurt(s_{j_0})z_{j_0}^2(k) + 3z_{j_0}^2(k))\right] z_{j_0}(k).$$

Give any $\tau > 0$, there exists a $K \geq 1$ such that

$$|\eta a - V(0)|\sqrt{\frac{1 + \eta a}{\eta a}} \frac{\zeta^K}{1 - \zeta} < \tau.$$

For any $k_1 > k_2 \geq K$, it follows that

$$
\begin{aligned}
|z_{j_0}(k_1) - z_{j_0}(k_2)| &= \left|\sum_{i=k_2}^{k_1-1}(z_{j_0}(i+1) - z_{j_0}(i))\right| \\
&= \left|\sum_{i=k_2}^{k_1-1}[(\eta a - V(i))z_{j_0}(i)]\right| \\
&\leq \sum_{i=k_2}^{k_1-1}|\eta a - V(i)||z_{j_0}(i)| \\
&\leq \sqrt{\frac{1 + \eta a}{\eta a}}\sum_{i=k_2}^{k_1-1}|\eta a - V(i)|.
\end{aligned}
$$

By Lemma 6.9, it holds that

$$
\begin{aligned}
|z_{j_0}(k_1) - z_{j_0}(k_2)| &\leq \sqrt{\frac{1 + \eta a}{\eta a}}\sum_{i=k_2}^{\infty}|\eta a - V(0)|\zeta^{i-1} \\
&\leq |\eta a - V(0)|\sqrt{\frac{1 + \eta a}{\eta a}}\frac{\zeta^{k_2-1}}{1 - \zeta} \\
&\leq \tau.
\end{aligned}
$$

This shows the sequence $\{z_{j_0}(k)\}$ is a Cauchy sequence. By Cauchy convergence principle, there must exist a $z_{j_0}^*$ so that $\lim\limits_{k \to +\infty} z_{j_0}(k) = z_{j_0}^*$. The proof is complete.

Theorem 6.3 shows the average evolution of the algorithm (6.1) converges to an independent component direction with a positive kurtosis. The analysis can be similarly used to study the convergence of (6.2). The results can be obtained as follows.

Theorem 6.4 *Suppose* $\mathbf{z}(0) \in S_{\bar{z}2}$ *and* $\mathbf{z}(0) \notin V_p^{\perp}$. *If* $-Kurt(s_{i_0})z_{i_0}^2(0) > -Kurt(s_i)z_i^2(0), i_0 \in P$, *and* $i \in T/i_0$, *there must exist a constant* $z_{j_0}^*$ *so that* $\lim\limits_{k \to +\infty} E\{\mathbf{w}(k)\} = \mathbf{b}_{j_0} \cdot z_{j_0}^*$, *where* $Kurt(s_{i_0}) < 0$.

The previous theorems show the trajectories of the DDT algorithms starting from the invariant sets will converge to an independent component direction with a positive kurtosis or a negative kurtosis. The results can shed some light on the dynamical behaviors of the original algorithms (6.1) and (6.2).

6.4 Extension of the DDT Algorithms

The DDT algorithms of the algorithms (6.1) and (6.2) can be clearly expressed as

$$E\{\mathbf{w}(k+1)\} = E\left\{\mathbf{w}(k) + \eta\left[b\mathbf{x}(k)(\mathbf{w}^T(k)\mathbf{x}(k))^3 + a(1 - \|\mathbf{w}(k)\|^2)\mathbf{w}(k)\right]\right\},$$
$$E\{\mathbf{w}(k+1)\} = E\left\{\mathbf{w}(k) + \eta\left[-b\mathbf{x}(k)(\mathbf{w}^T(k)\mathbf{x}(k))^3 + a(1 - \|\mathbf{w}(k)\|^2)\mathbf{w}(k)\right]\right\}.$$

Let $\bar{\mathbf{w}}(k) = E\{\mathbf{w}(k)\}$, the DDT algorithms are rewritten as:

$$\bar{\mathbf{w}}(k+1) = \bar{\mathbf{w}}(k) + \eta\left[bE\{\mathbf{x}(k)(\bar{\mathbf{w}}^T(k)\mathbf{x}(k))^3\} + a(1 - \|\bar{\mathbf{w}}(k)\|^2)\bar{\mathbf{w}}(k)\right],$$
(6.10)

$$\bar{\mathbf{w}}(k+1) = \bar{\mathbf{w}}(k) + \eta\left[-bE\{\mathbf{x}(k)(\bar{\mathbf{w}}^T(k)\mathbf{x}(k))^3\} + a(1 - \|\bar{\mathbf{w}}(k)\|^2)\bar{\mathbf{w}}(k)\right].$$
(6.11)

By studying the convergence of the DDT algorithms (6.10) and (6.11), we indirectly analyze the convergence of the original SDT algorithms (6.1) and (6.2). Clearly, the algorithms will converge under certain conditions. However, the original SDT algorithms have slow convergence speed, and convergence accuracy of these algorithms is not good [129, 197, 198]. To improve the performance of the original algorithms, the DDT algorithms can be extended to the block versions of the original algorithms. If the expectation operation is computed by

$$E\{\mathbf{x}(k)(\bar{\mathbf{w}}^T(k)\mathbf{x}(k))^3\} = \frac{1}{L}\sum_{j=1}^{L}\mathbf{x}(j)(\bar{\mathbf{w}}^T(j)\mathbf{x}(j))^3,$$

where L is the block size [129], the algorithms (6.10) and (6.11) are also the block algorithms. The block algorithms establish a relationship between the original SDT algorithms and the corresponding DDT algorithms. Clearly, if $L = 1$, they are the original SDT algorithms. If L is the size of all examples, they are the corresponding DDT algorithms. The block algorithms represent an average evolution of the original algorithms over all blocks of samples. Thus, the invariant sets in Theorem 6.1 (where $L = 1$) also are the invariant sets of the block algorithms. Furthermore, the block algorithms can get a good convergence speed and accuracy in practice. The following simulations will illustrate their performance.

6.5   Simulation and Discussion

In this section, three set of experiments will be carried out to illustrate the convergence of the class of Hyvärinen-Oja's ICA learning algorithms with constant learning rates.

6.5.1   Example 1

The DDT algorithms characterize the average evolution of the original algorithms. The convergence of the DDT algorithms can reflect the dynamical behaviors of the original algorithms in the invariant sets. From (6.7), consider four-dimensional DDT algorithms as

$$
\begin{bmatrix} z_1(k+1) \\ z_2(k+1) \\ z_3(k+1) \\ z_4(k+1) \end{bmatrix} = \begin{bmatrix} z_1(k) \\ z_2(k) \\ z_3(k) \\ z_4(k) \end{bmatrix} + \eta \left\{ \sigma b \begin{bmatrix} Kurt(s_1)z_1^3(k) \\ Kurt(s_2)z_2^3(k) \\ Kurt(s_3)z_3^3(k) \\ Kurt(s_4)z_4^3(k) \end{bmatrix} \right.
$$
$$
\left. + \left(3\sigma b \|\mathbf{z}(k)\|^2 + \eta a (1 - \|\mathbf{z}(k)\|^2)\right) \begin{bmatrix} z_1(k) \\ z_2(k) \\ z_3(k) \\ z_4(k) \end{bmatrix} \right\},
$$

for all $k \geq 0$. Let $\eta = 0.1, a = 5$, and $b = 0.5$. While $\sigma = 1$ and $Kurt(s_1) = -0.6, Kurt(s_2) = 0, Kurt(s_3) = 3, Kurt(s_4) = 0.4$, in Figure 6.1, the evolution of components of $\mathbf{z}$ is shown with initial vector $\mathbf{z}(0) = [-0.3 \quad -0.6 \quad 0.2 \quad 0.4]^T$ (upper) and $\mathbf{z}(0) = [-1 \quad -0.5 \quad -0.1 \quad -0.2]^T$ (bottom). While $\sigma = -1$ and the initial value $\mathbf{z}(0) = [-0.4 \quad -0.8 \quad 0.2 \quad 0.4]^T$, Figure 6.2 shows the evolution results of components of $\mathbf{z}$ with the kurtosis $Kurt(s_1) = -0.6, Kurt(s_2) = 0, Kurt(s_3) = 3, Kurt(s_4) = -0.4$ (upper) and $Kurt(s_1) = -0.6, Kurt(s_2) = 0, Kurt(s_3) = -2.5, Kurt(s_4) = -0.4$ (bottom).

Since $E\{\mathbf{w}(k)\} = B\mathbf{z}(k)$, the DDT algorithms converge to an independent component direction with a positive kurtosis or a negative kurtosis as all components of $\mathbf{z}(k)$ converge to zero except a component. The simulation illustrates the results of Theorems 6.3 and 6.4.

In addition, the initial value and kurtosis together determine which independent component will be extracted in the DDT algorithms. However, in a stochastic environment, the distribution of the variable also affects which independent component will be extracted. The following experiments will further illustrate the theory derived.

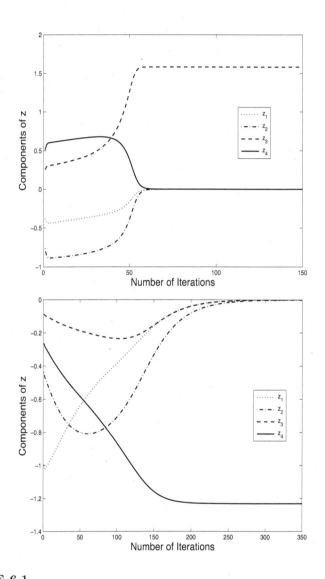

FIGURE 6.1

The evolution of components of **z** with initial value: $\mathbf{z}(0) = [-0.3 \quad -0.6 \quad 0.2 \quad 0.4]^T$ (upper) and $\mathbf{z}(0) = [-1 \quad -0.5 \quad -0.1 \quad -0.2]^T$ (bottom), while $\sigma = 1$ and $Kurt(s_1) = -0.6, Kurt(s_2) = 0, Kurt(s_3) = 3, Kurt(S_4) = 0.4$.

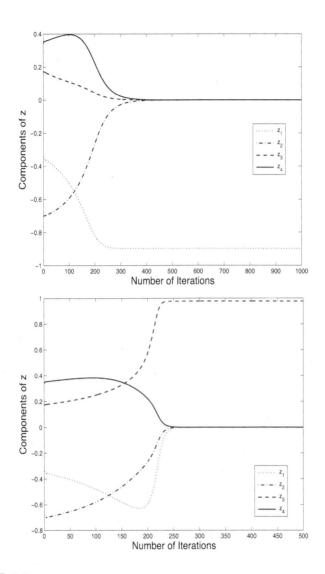

FIGURE 6.2

The evolution of components of $\mathbf{z}$ with the different kurtosis: $Kurt(s_1) = -0.6, Kurt(s_2) = 0, Kurt(s_3) = 3, Kurt(S_4) = -0.4$ (upper); $Kurt(s_1) = -0.6, Kurt(s_2) = 0, Kurt(s_3) = -2.5, Kurt(S_4) = -0.4$ (bottom), while $\sigma = -1$ and the initial value: $\mathbf{z}(0) = [-0.4 \quad -0.8 \quad 0.2 \quad 0.4]^T$.

6.5.2   Example 2

The example will illustrate the convergence of the block algorithm (6.10). There are two stochastic signals s_1 and s_2 with $Kurt(s_1) = 2.6822, Kurt(s_2) = 2.4146$. The original signals are artificially generated and are samples of 1000 values with zero mean and unit variance, shown in Figure 6.3 (upper). Randomly mixed and written signals are shown in Figure 6.3 (bottom). Since $\Gamma_w = 1192.7$, let $\eta = 0.05, a = 25$, and $b = 0.01$. The algorithm (6.10) is used to extract an independent signal from the mixed signals. The evolution of norm of $\mathbf{w}$ is shown in Figure 6.4 with $L = 200, \mathbf{w}(0) = [0.03, 0, 5]$ (upper) and $L = 1000, \mathbf{w}(0) = [0.3, 0.05]$ (bottom). Clearly, the norm changes in a certain range and will not go to infinity.

To verify whether the direction of $\mathbf{w}$ converges, we will compute the *DirectionCosine* between $\mathbf{w}$ and a reference vector $\mathbf{v} = [1 \ 1]^T$ at each k as:

$$DirectionCosine(k) = \frac{|\mathbf{w}^T(k) \cdot \mathbf{v}|}{\|\mathbf{w}(k)\| \cdot \|\mathbf{v}\|},$$

for all $k \geq 0$. We do not known the exact extracting direction and which signal will be extracted in advance. Thus, a reference vector is selected. Clearly, *DirectionCosine(k)* will approach to a certain value if the direction of $\mathbf{w}$ converges. Figure 6.5 shows the evolution of *Directioncosine* with $L = 200, \mathbf{w}(0) = [0.03, 0, 5]$ (upper) and $L = 1000, \mathbf{w}(0) = [0.3, 0.05]$ (bottom).

At the same time, to verify whether the direction that $\mathbf{w}$ goes to is right, a signal is extracted by projecting the mixed whitened signals to the direction. The extracted signals are shown in Figure 6.6. The example shows that the block algorithm (6.10) exists an invariant set and converges to an independent component direction in the invariant set.

6.5.3   Example 3

In the following experiment, four signals are used, as shown in Figure 6.7 (upper). They are of zero mean and unit variance with $Kurt(s_1) = -1.4984, Kurt(s_2) = 0.0362, Kurt(s_3) = -1.1995$, and $Kurt(s_4) = 2.4681$. Since $\Gamma_w = 471.8463$, let $\eta = 0.05, a = 12$, and $b = 0.01$. The original SDT algorithms (6.1) and (6.2) are used to extract the independent signals from the mixed and written signals, as shown in Figure 6.7 (bottom). As for the algorithm (6.1), the evolution of norm and *DirectionCosine* are shown in Figure 6.8. An independent signal with a positive kurtosis is extracted, shown in Figure 6.10 (upper). As for the algorithm (6.2), the evolution of norm and *DirectionCosine* are shown in Figure 6.9. An independent signal with a negative kurtosis is extracted, as shown in Figure 6.10 (bottom).

We also compare the convergence rate of the algorithms (6.1) and (6.2) with a constant learning rate $\eta = 0.05$ and $\eta(k) = \frac{0.1}{2log_2(k+1)}, (k \geq 1)$. The evolution trajectories of *DirectionCosine* are shown in Figures 6.11 and 6.12.

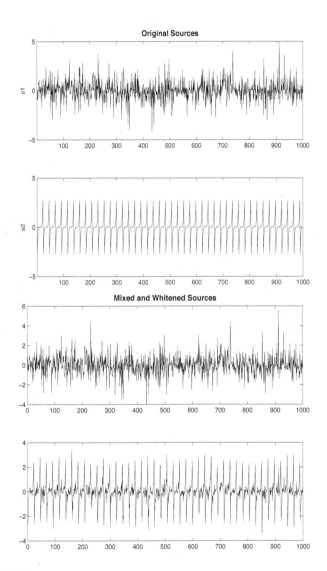

FIGURE 6.3
The original sources s_1 and s_2 with $Kurt(s_1) = 2.6822, Kurt(s_2) = 2.4146$ (upper); two mixed and whitened signals (bottom).

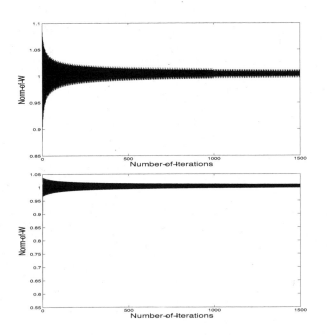

FIGURE 6.4

The evolution of norm of $\mathbf{w}$ with $L = 200, \mathbf{w}(0) = [0.3 \quad 0.05]^T$ (upper) and with $L = 1000, \mathbf{w}(0) = [0.03 \quad 0.5]^T$ (bottom).

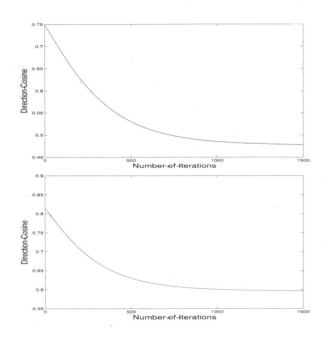

FIGURE 6.5
The evolution of *DirectionCosine(k)* with $L = 200, \mathbf{w}(0) = [0.3 \ 0.05]^T$ (upper) and with $L = 1000, \mathbf{w}(0) = [0.03 \ 0.5]^T$ (bottom).

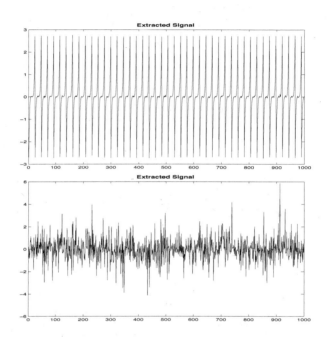

FIGURE 6.6
The extracted signal with $L = 200, \mathbf{w}(0) = [0.3 \quad 0.05]^T$ (upper) and with $L = 1000, \mathbf{w}(0) = [0.03 \quad 0.5]^T$ (bottom).

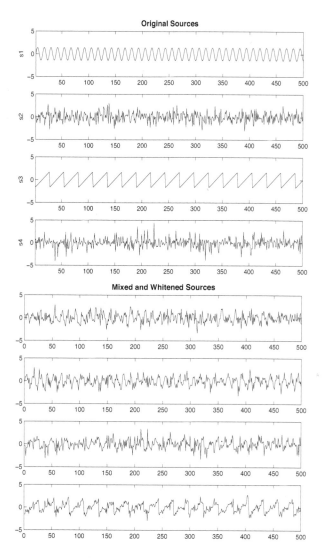

FIGURE 6.7
The original sources s_1, s_2, s_3, and s_4 with $Kurt(s_1) = -1.4984, Kurt(s_2) = 0.0362, Kurt(s_3) = -1.1995$ $Kurt(s_4) = 2.4681$ (upper); four mixed and whitened signals (bottom).

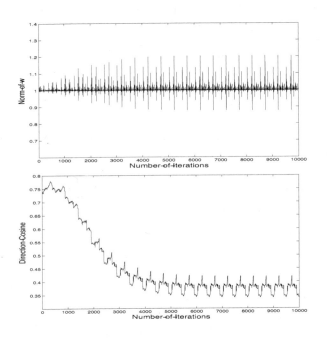

FIGURE 6.8

The evolution of norm of **w** (upper) and the evolution of *DirectionCo-sine(k)* (bottom) of the algorithm (6.1) with the initial value $\mathbf{w}(0) = [0.6 \quad 1.2 \quad 0.1 \quad 0.1]^T$.

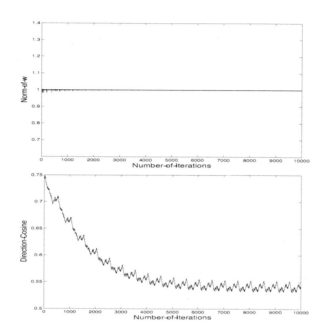

FIGURE 6.9

The evolution of norm of **w** (upper) and the evolution of *DirectionCosine(k)* (bottom) of the algorithm (6.2) with the initial value **w**(0) = $[0.6 \quad 1.2 \quad 0.1 \quad 0.1]^T$.

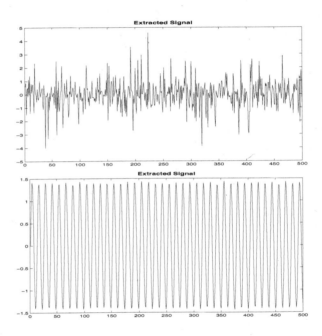

FIGURE 6.10
The extracted signal of (6.1) with $Kurt(s_4) = 2.4681$ (upper) and the extracted signal of (6.2) with $Kurt(s_1) = -1.4984$ (bottom).

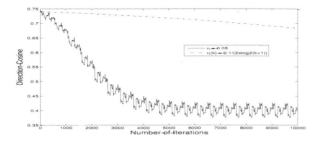

FIGURE 6.11

Comparison of convergent rate of algorithm (6.1) with a constant learning rate $\eta = 0.05$ and a zero-approaching learning rate $\eta(k) = \frac{0.1}{2log_2(k+1)}$. The dash-dot line denotes the evolution of the algorithm with the a zero-approaching learning rate.

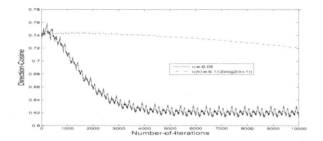

FIGURE 6.12

Comparison of convergent rate of algorithm (6.2) with a constant learning rate $\eta = 0.05$ and a zero-approaching learning rate $\eta(k) = \frac{0.1}{2log_2(k+1)}$. The dash-dot line denotes the evolution of the algorithm with the a zero-approaching learning rate.

The dash-dot line denotes the evolution of the algorithms with the zero-approaching learning rate. Clearly, the algorithms with a constant learning rate converge faster in this experiment. Maybe it is possible to improve the performance by selecting a proper zero-approaching learning rate. However, there is no idea of selecting a proper zero-approaching learning rate so that the algorithms do not diverge and converge faster. Constant learning rates simplify the application of the original algorithms in practice and the convergent properties can be guaranteed in stochastic environment.

6.6   Conclusion

The convergence of a class of Hyvärinen-Oja's ICA learning algorithms with constant learning algorithms is studied. It is shown that the algorithms converge to an independent component direction with a positive kurtosis or a negative kurtosis. Since constant learning rates are used, the requirements for the algorithms to be applied in practical applications can be met. The most important result of this chapter is that some invariant sets are obtained in stochastic environment so that the nondivergence of the original SDT algorithms is guaranteed by selecting proper learning parameters. In these invariant sets, the local convergence of the original algorithms has been indirectly studied via the DDT method. To improve the performance of the original algorithms, the corresponding DDT algorithms can be extended to the block versions of the original algorithms. Simulations are carried out to further support the results obtained.

7

Chaotic Behaviors Arising from Learning Algorithms

7.1 Introduction

In the previous chapters, we have discussed the convergence of some subspace learning algorithms by mainly using the DDT method. Some invariant sets of these algorithms will be obtained under certain conditions. The convergence conditions of these algorithms will be derived. However, some learning algorithms exhibit some chaotic behaviors in practical applications. In [123], Xu's LMSER algorithm can produce chaotic behaviors when the learning rate is chosen in certain range. The article [122] also explores chaotic behaviors of the Douglas's MCA learning algorithm [60].

Chaotic behaviors are crucial to use learning algorithms in practical applications. On the one hand, chaotic behavior may result in the breakdown of the system [178]. Thus, how to choose a system parameter to avoid the chaotic behavior is very important. See examples [2, 78, 79, 92, 110]. On the other hand, chaotic behaviors can be applied in many scientific disciplines: mathematics, finance [105], biology [161, 6], engineering [34, 33, 140, 163, 183], and ecology [58, 16].

In this chapter, we explore the stability and chaotic behavior of a class of ICA algorithms of the one-dimensional type [121]. Though we only discuss the one-dimensional case of the original algorithms, this can shed light on the dynamical behavior of the original algorithms.

In [86], Hyvärinen and Oja proposed some neuron models for ICA. The corresponding learning algorithms are described as

$$\mathbf{w}(k + 1) = \mathbf{w}(k) + \mu(k)[-\sigma\mathbf{x}(k)g(\mathbf{w}^T(k)\mathbf{x}(k)) - f(\|\mathbf{w}(k)\|^2)\mathbf{w}(k)],$$

where f is a scalar function, and $\sigma = \pm 1$ is a sign that determines whether we are minimizing or maximizing kurtosis [86]. $\mu(k)$ is the learning rate sequence and $\mathbf{x}(k)$ is a prewhiten input data. The function g is a polynomial:

$$g(y) = ay - by^3,$$

where $a \geq 0$ and $b > 0$.

In practical applications, constant learning rates are usually required to speed up the convergence and satisfy the computational requirements [125, 127, 185, 210]. In this chapter, the dynamical behaviors of the algorithms mentioned above with constant learning rates are studied. These algorithms with constant learning rates in the one-dimensional case can be presented as

$$w(k+1) = w(k) + \eta[-\sigma g(w^T(k)) - f(\|w(k)\|^2)w(k)],$$

for all $k \geq 0$, where f is a scalar function, $\eta > 0$ is a constant learning rate. Clearly, if $\sigma = 1$, the algorithms can be rewritten as:

$$w(k+1) = w(k) + \eta[-(aw(k) - bw^3(k)) - f(\|w(k)\|^2)w(k)]. \quad (7.1)$$

If $\sigma = -1$, it follows that

$$w(k+1) = w(k) + \eta[(aw(k) - bw^3(k)) - f(\|w(k)\|^2)w(k)]. \quad (7.2)$$

This chapter will show the algorithms (7.1) and (7.2) have both stability and chaotic behaviors under some conditions. These conditions for stability and chaos will be derived. Therefore, it is easy to see that divergent and chaotic behaviors can be avoided by choosing proper learning parameters.

This chapter is organized as follows. Section 2 identifies some invariant sets and obtains the conditions for divergence of these algorithms. In Section 3, the conditions for the stability of these algorithms are derived. The chaos and bifurcation of the algorithms are analyzed in Section 4. Finally, conclusions are drawn in Section 5.

7.2   Invariant Set and Divergence

Clearly, an invariant set guarantees the nondivergence of an algorithm. The following theorem will give the invariant sets of the algorithms (7.1) and (7.2). For convenience of analysis, two lemmas are given first.

Denote by

$$M_1^* = max \left\{ \frac{a + f(\|w(k)\|^2)}{b}, \frac{4(\eta(a + f(\|w(k)\|^2)) - 1)^3}{27\eta b} \right\},$$

$$M_2^* = max \left\{ \frac{\eta(a - f(\|w(k)\|^2)) + 2}{\eta b}, \frac{4(\eta(a - f(\|w(k)\|^2)) + 1)^3}{27\eta b} \right\}.$$

Lemma 7.1 *Suppose that $\eta > 0$, $a > 0$, and $b > 0$. It holds that*

$$\left[1 - \eta a - \eta f(\|w(k)\|^2) + \eta bs\right]^2 \cdot s \leq M_1^*,$$

for all $0 < s \leq \dfrac{a + f(\|w(k)\|^2)}{b}$ and $\eta(a + f(\|w(k)\|^2)) > 0$.

Proof: Define a differentiable function

$$f(s) = \left[1 - \eta a - \eta f(\|w(k)\|^2) + \eta bs\right]^2 \cdot s,$$

for all $0 < s \leq \dfrac{a + f(\|w(k)\|^2)}{b}$. It follows that

$$\dot{f}(s) = \left[1 - \eta a - \eta f(\|w(k)\|^2) + \eta bs\right]\left[1 - \eta a - \eta f(\|w(k)\|^2) + 3\eta bs\right],$$

for all $0 < s \leq \dfrac{a + f(\|w(k)\|^2)}{b}$. Denote

$$\xi_1 = \frac{\eta(a + f(\|w(k)\|^2)) - 1}{\eta b}, \quad \xi_2 = \frac{\eta(a + f(\|w(k)\|^2)) - 1}{3\eta b}.$$

If $\eta(a + f(\|w(k)\|^2)) > 1$, it follows that

$$\dot{f}(s) \begin{cases} > 0, & \text{if } 0 < s < \xi_1 \\ = 0, & \text{if } s = \xi_1 \\ < 0, & \text{if } \xi_1 < s < \xi_2 \\ = 0, & \text{if } s = \xi_2 \\ > 0, & \text{if } s > \xi_2 \end{cases} \quad .$$

This shows that the maximum point of the function $f(s)$ on the interval $\left(0, \dfrac{a + f(\|w(k)\|^2)}{b}\right)$ must be $s^\star = \xi_1$ or $s^\star = \dfrac{a + f(\|w(k)\|^2)}{b}$. If $0 < \eta(a + f(\|w(k)\|^2)) \leq 1$, then $s^\star = \dfrac{a + f(\|w(k)\|^2)}{b}$ must be the maximum point of $f(s)$ on the interval $\left(0, \dfrac{a + f(\|w(k)\|^2)}{b}\right)$. Thus, it is not difficult to get that

$$\left[1 - \eta a - \eta f(\|w(k)\|^2) + \eta bs\right]^2 \cdot s \leq M_1^\star,$$

for all $0 < s \leq \dfrac{a + f(\|w(k)\|^2)}{b}$. The proof is completed.

Lemma 7.2 *Suppose that $\eta > 0$, $a > 0$, and $b > 0$. It holds that*

$$\left[1 + \eta a - \eta f(\|w(k)\|^2) - \eta bs\right]^2 \cdot s \leq M_2^\star,$$

for all $0 < s \leq \dfrac{\eta(a - f(\|w(k)\|^2)) + 2}{\eta b}$ and $\eta(a - f(\|w(k)\|^2)) > -2$.

Proof: Define a differentiable function

$$f(s) = \left[1 + \eta a - \eta f(\|w(k)\|^2) - \eta bs\right]^2 \cdot s$$

for all $0 < s \leq \dfrac{\eta(a - f(\|w(k)\|^2)) + 2}{\eta b}$. It follows that

$$\dot{f}(s) = \left[1 + \eta a - \eta f(\|w(k)\|^2) - \eta bs\right] \left[1 + \eta a - \eta f(\|w(k)\|^2) - 3\eta bs\right],$$

for all $0 < s \leq \dfrac{\eta(a - f(\|w(k)\|^2)) + 2}{\eta b}$. Denote

$$\xi_1 = \frac{\eta(a - f(\|w(k)\|^2)) + 1}{\eta b}, \quad \xi_2 = \frac{\eta(a - f(\|w(k)\|^2)) + 1}{3\eta b}.$$

If $\eta(a - f(\|w(k)\|^2)) > -1$, it follows that

$$\dot{f}(s) \begin{cases} > 0, & \text{if } 0 < s < \xi_1 \\ = 0, & \text{if } s = \xi_1 \\ < 0, & \text{if } \xi_1 < s < \xi_2 \\ = 0, & \text{if } s = \xi_2 \\ > 0, & \text{if } s > \xi_2 \end{cases}.$$

This shows that the maximum point of the function $f(s)$ on the interval $\left(0, \dfrac{\eta(a - f(\|w(k)\|^2)) + 2}{\eta b}\right)$ must be $s^\star = \xi_1$ or $s^\star = \dfrac{\eta(a - f(\|w(k)\|^2)) + 2}{\eta b}$.

If $-2 < \eta(a - f(\|w(k)\|^2)) \leq -1$, $s^\star = \dfrac{\eta(a - f(\|w(k)\|^2)) + 2}{\eta b}$ must be the maximum point of $f(s)$. Thus, it is not difficult to get that

$$\left[1 + \eta a - \eta f(\|w(k)\|^2) - \eta bs\right]^2 \cdot s \leq M_2^\star,$$

for all $0 < s \leq \dfrac{\eta(a - f(\|w(k)\|^2)) + 2}{\eta b}$. The proof is completed.

Lemma 7.3 *If $0 < x \leq 4$, it holds that $(x - 1)^3 \leq \dfrac{27x}{4}$.*

Proof: Define a differentiable function

$$f(x) = (x - 1)^3 - \frac{27x}{4},$$

for all $0 < x \leq 4$. It follows that

$$\dot{f}(x) = 3(x - 1)^2 - \frac{27}{4},$$

for all $0 < x \leq 4$. Clearly, if $0 < x \leq \dfrac{5}{2}$, it follows that $\dot{f}(x) \leq 0$. Thus, it holds that $f(x) \leq f(0) = -1$ if $0 < x \leq \dfrac{5}{2}$. If $\dfrac{5}{2} < x \leq 4$, it follows that $\dot{f}(x) > 0$. Thus, it holds that $f(x) \leq f(4) = 0$ if $\dfrac{5}{2} < x \leq 4$. So, it is easy to see that $f(x) \leq 0$ if $0 < x \leq 4$. The proof is completed.

Denote

$$S_1 = \left\{ w^2 | w^2 \in R, w^2 < \frac{a + f(\|w(k)\|^2)}{b} \right\},$$

$$S_2 = \left\{ w^2 | w^2 \in R, w^2 < \frac{a - f(\|w(k)\|^2)}{b} + \frac{2}{\eta b} \right\}.$$

Theorem 7.1 *Suppose $\eta > 0$, $a > 0$, and $b > 0$. If $0 < \eta(a + f(\|w(k)\|^2)) \leq 4$, then S_1 is an invariant set of (7.1). If $-2 < \eta(a - f(\|w(k)\|^2)) \leq 2$, then S_2 is an invariant set of (7.2).*

Proof : From (7.1), it follows that

$$w^2(k+1) = \left[1 - \eta a - \eta f(\|w(k)\|^2) + \eta b w^2(k) \right]^2 \cdot w^2(k),$$

for all $k \geq 0$. So, it holds that

$$
\begin{aligned}
w^2(k+1) &\leq \max \left\{ \left[1 - \eta a - \eta f(\|w(k)\|^2) + \eta b w^2(k) \right]^2 \cdot w^2(k) \right\} \\
&\leq \max \left\{ \left[1 - \eta a - \eta f(\|w(k)\|^2) + \eta b s \right]^2 \cdot s \right\},
\end{aligned}
$$

for all $0 < s \leq \dfrac{a + f(\|w(k)\|^2)}{b}$. By Lemma 7.1, it follows that $w^2(k+1) \leq M_1^\star$, for $k \geq 0$. By Lemma 7.3, if $0 < \eta(a + f(\|w(k)\|^2)) \leq 4$, it holds that

$$\frac{4(\eta(a + f(\|w(k)\|^2)) - 1)^3}{27\eta b} \leq \frac{a + f(\|w(k)\|^2)}{b}.$$

So, it follows that

$$w^2(k+1) \leq \frac{a + f(\|w(k)\|^2)}{b}.$$

Thus, S_1 is an invariant set of (7.1).

From (7.2), it follows that

$$w^2(k+1) = \left[1 + \eta a - \eta f(\|w(k)\|^2) - \eta b w^2(k) \right]^2 \cdot w^2(k),$$

for all $k \geq 0$. Thus, it holds that

$$
\begin{aligned}
w^2(k+1) &\leq \max \left\{ \left[1 + \eta a - \eta f(\|w(k)\|^2) - \eta b w^2(k) \right]^2 \cdot w^2(k) \right\} \\
&\leq \max \left\{ \left[1 + \eta a - \eta f(\|w(k)\|^2) - \eta b s \right]^2 \cdot s \right\},
\end{aligned}
$$

for all $0 < s \leq \dfrac{\eta(a - f(\|w(k)\|^2)) + 2}{\eta b}$. By Lemma 7.2, it follows that $w^2(k+1) \leq M_2^\star$, for $k \geq 0$. By Lemma 7.3, if $-2 < \eta(a - f(\|w(k)\|^2)) \leq 2$, it holds that

$$\frac{4(\eta(a - f(\|w(k)\|^2)) + 1)^3}{27\eta b} \leq \frac{\eta(a - f(\|w(k)\|^2)) + 2}{\eta b}.$$

So, if follows that

$$w^2(k+1) \leq \frac{\eta(a - f(\|w(k)\|^2)) + 2}{\eta b}.$$

Thus, S_2 is an invariant set of (7.2). The proof is completed.

The previous theorem shows any trajectory of algorithms (7.1) or (7.2) starting from $w(0)$ in the invariant sets S_1 or S_2 will remain in S_1 or S_2. This guarantees the nondivergence of the algorithms (7.1) and (7.2). However, outside of the invariant sets, the algorithms exhibit different dynamical behaviors, even possibly diverging to infinity. The following theorems give the divergent conditions of the algorithms.

Theorem 7.2 *Suppose that $\eta > 0$, $a > 0$, $b > 0$, and $w(0) \neq 0$. For the algorithm (7.1), if $0 < \eta(a + f(\|w(k)\|^2))$ and*

$$w^2(k) > \frac{a + f(\|w(k)\|^2)}{b},$$

then $\lim_{k \to +\infty} w^2(k+1) = +\infty$. For the algorithm (7.2), if $-2 < \eta(a - f(\|w(k)\|^2))$ and

$$w^2(k) > \frac{\eta(a - f(\|w(k)\|^2)) + 2}{\eta b},$$

then $\lim_{k \to +\infty} w^2(k+1) = +\infty$.

Proof : From (7.1), it holds that

$$w^2(k+1) = \left[1 - \eta a - \eta f(\|w(k)\|^2) + \eta b w^2(k)\right]^2 \cdot w^2(k),$$

for all $k \geq 0$. If

$$w^2(k) > \frac{a + f(\|w(k)\|^2)}{b},$$

for all $k \geq 0$, it holds that

$$1 - \eta a - \eta f(\|w(k)\|^2) + \eta b w^2(k) > 1,$$

for all $k \geq 0$. Then we have

$$\left[1 - \eta a - \eta f(\|w(k)\|^2) + \eta b w^2(k)\right]^2 > 1,$$

for all $k \geq 0$. Thus, for the algorithm (7.1), it holds that $\lim_{k \to +\infty} w^2(k+1) = +\infty$.

From the algorithm (7.2), it follows that

$$w^2(k+1) = \left[1 + \eta a - \eta f(\|w(k)\|^2) - \eta b w^2(k)\right]^2 \cdot w^2(k),$$

for all $k \geq 0$. If

$$w^2(k) > \frac{\eta(a - f(\|w(k)\|^2)) + 2}{\eta b},$$

for all $k \geq 0$, it holds that

$$1 + \eta a - \eta f(\|w(k)\|^2) - \eta b w^2(k) < -1,$$

for all $k \geq 0$. It is easy to see that

$$\left[1 + \eta a - \eta f(\|w(k)\|^2) - \eta b w^2(k)\right]^2 > 1,$$

for all $k \geq 0$. Thus, $\lim\limits_{k \to +\infty} w^2(k+1) = +\infty$ for the algorithm (7.2).
The proof is completed.

Theorem 7.3 *Suppose that $\eta > 0$, $a > 0$, $b > 0$, and $w(0) \neq 0$. For the algorithm (7.1), if $\eta(a + f(\|w(k)\|^2)) \leq 0$, then $\lim\limits_{k \to +\infty} w^2(k+1) = +\infty$. For the algorithm (7.2), if $\eta(a - f(\|w(k)\|^2)) \leq -2$, then $\lim\limits_{k \to +\infty} w^2(k+1) = +\infty$.*

Proof: From (7.1), it holds that

$$w^2(k+1) = \left[1 - \eta a - \eta f(\|w(k)\|^2) + \eta b w^2(k)\right]^2 \cdot w^2(k),$$

for all $k \geq 0$. If $\eta(a + f(\|w(k)\|^2)) \leq 0$ and $w(0) \neq 0$, it holds that

$$\left[1 - \eta a - \eta f(\|w(k)\|^2) + \eta b w^2(k)\right]^2 > 1,$$

for all $k \geq 0$. Thus, for the algorithm (7.1), it holds that $\lim\limits_{k \to +\infty} w^2(k+1) = +\infty$.

From the algorithm (7.2), it follows that

$$w^2(k+1) = \left[1 + \eta a - \eta f(\|w(k)\|^2) - \eta b w^2(k)\right]^2 \cdot w^2(k),$$

for all $k \geq 0$. If $\eta(a - f(\|w(k)\|^2)) < -2$ and $w(0) \neq 0$, it holds that

$$\left[1 + \eta a - \eta f(\|w(k)\|^2) - \eta b w^2(k)\right]^2 > 1,$$

for all $k \geq 0$. Thus, $\lim\limits_{k \to +\infty} w^2(k+1) = +\infty$ for the algorithm (7.2).
The proof is completed.

From the previous theorems, the invariant sets and divergent regions are shown in Figures 7.1 and 7.2. The S_1 and S_2 are the invariant sets of the algorithms (7.1) and (7.2), respectively. The regions I, II, and III are divergent regions. In the IV regions, the bandedness of the algorithms cannot be guaranteed. The following example (Figure 7.3) illustrates the divergence of (7.1) and (7.2) in region IV.

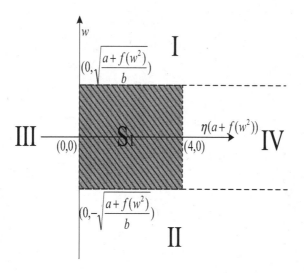

FIGURE 7.1
The invariant set S_1 of the algorithm (7.1), and the divergent regions I, II, and III.

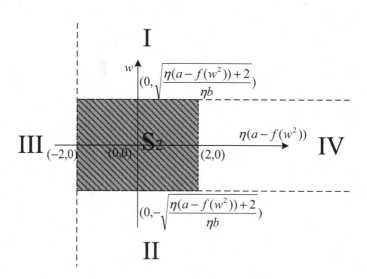

FIGURE 7.2
The invariant set S_2 of the algorithm (7.2), and the divergent regions I, II, and III.

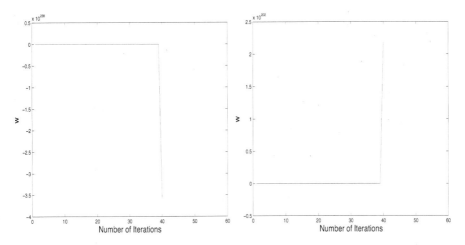

FIGURE 7.3
Divergence of the algorithm (7.1) with $\eta(a + f(\|w(k)\|^2))) = 4.01$ and $w^2(0) = 0.01 < \frac{(a + f(\|w(k)\|^2)))}{b} = 2.6733$ (left), divergence of the algorithm (7.2) with $\eta(a - f(\|w(k)\|^2))) = 2.01$ and $w^2(0) = 0.01 < \frac{\eta(a - f(\|w(k)\|^2)) + 2}{\eta b} = 2.6733$ (right).

7.3 Stability Analysis

In this section, we will explore the stability of the algorithms (7.1) and (7.2) in invariant set S_1 and S_2.

Definition 7.1 *A point $w^* \in R^n$ is called an equilibrium of (7.1), if and only if*

$$w^* = w^* + \eta \left[-(aw^* - b(w^*)^3) - f(\|w(k)\|^2)w^* \right].$$

Definition 7.2 *A point $w^* \in R^n$ is called an equilibrium of (7.2), if and only if*

$$w^* = w^* - \eta \left[aw^* - b(w^*)^3 - f(\|w(k)\|^2)w^* \right].$$

Clearly, the set of all equilibrium points of (7.1) is

$$\left\{ 0, \sqrt{\frac{a + f(\|w(k)\|^2)}{b}}, -\sqrt{\frac{a + f(\|w(k)\|^2)}{b}} \right\}.$$

The set of all equilibrium points of (7.2) is

$$\left\{ 0, \sqrt{\frac{a - f(\|w(k)\|^2)}{b}}, -\sqrt{\frac{a - f(\|w(k)\|^2)}{b}} \right\}.$$

Obviously, the algorithms have many equilibria, thus, the study of dynamics belongs to a multistability problem [192].

According to the Lyapunov indirect method, an equilibrium of an algorithm is stable if the absolute of each eigenvalue of the Jacobian matrix of the algorithm at this point is less than 1 [2]. For the algorithms (7.1) and (7.2), we will compute the eigenvalues of Jacobian matrix at each equilibrium point.

Theorem 7.4 *For the algorithm (7.1), in the invariant set S_1, the equilibrium 0 is stable if $0 < \eta(a + f(\|w(k)\|^2)) < 2$. $\pm\sqrt{\frac{a+f(\|w(k)\|^2)}{b}}$ are unstable points if $\eta(a + f(\|w(k)\|^2)) > 0$.*

Proof : Let

$$G = w(k) + \eta[-(aw(k) - bw^3(k)) - f(\|w(k)\|^2)w(k)].$$

We have

$$\frac{dG}{dw(k)} = 1 - \eta a - \eta f(\|w(k)\|^2) + 3\eta bw^2(k)).$$

Clearly, $\left.\frac{dG}{dw(k)}\right|_0 = 1 - \eta a - \eta f(\|w(k)\|^2)$. If $0 < \eta(a + f(\|w(k)\|^2)) < 2$, it holds that $\left|\frac{dG}{dw(k)}\right| < 1$. Thus, 0 is a stable point. As for equilibria $\pm\sqrt{\frac{a+f(\|w(k)\|^2)}{b}}$, $\left|\frac{dG}{dw(k)}\right| = 1 + 2\eta(a + f(\|w(k)\|^2))b)$. If $\eta(a + f(\|w(k)\|^2))b) > 0$, $\left|\frac{dG}{dw(k)}\right| > 1$. Thus, $\pm\sqrt{\frac{a+f(\|w(k)\|^2)}{b}}$ are unstable points for $\eta(a + f(\|w(k)\|^2))b) > 0$. The proof is completed.

Theorem 7.4 shows that in invariant set S_1, the algorithm (7.1) converges to 0 if $0 < \eta(a + f(\|w(k)\|^2))b) < 2$, as shown in Figure 7.4.

Theorem 7.5 *For the algorithm (7.2), in the invariant set S_2, the equilibrium 0 is stable if $-2 < \eta(a + f(\|w(k)\|^2)) \leq 0$. $\pm\sqrt{\frac{a-f(\|w(k)\|^2)}{b}}$ are stable points if $0 < \eta(a + f(\|w(k)\|^2))b) < 1$.*

Proof : Let

$$G = w(k) + \eta[(aw(k) - bw^3(k)) - f(\|w(k)\|^2)w(k)].$$

We have

$$\frac{dG}{dw(k)} = 1 + \eta a - \eta f(\|w(k)\|^2) - 3\eta bw^2(k)).$$

Clearly, $\left.\frac{dG}{dw(k)}\right|_0 = 1 + \eta a - \eta f(\|w(k)\|^2)$. If $-2 < \eta(a + f(\|w(k)\|^2)) < 0$, it holds that $\left|\frac{dG}{dw(k)}\right| < 1$. Thus, 0 is stable for $-2 < \eta(a + f(\|w(k)\|^2)) < 0$.

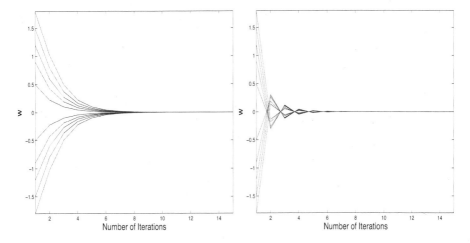

FIGURE 7.4
The algorithm (7.1) converges to 0 with the different initial values and with $b = 0.2, \eta(a + f) = 0.56$ (left) and $\eta(a + f) = 1.4$ (right).

As for equilibria $\pm\sqrt{\frac{a - f(\|w(k)\|^2)}{b}}$, $\left|\frac{dG}{dw(k)}\right| = 1 - 2\eta(a - f(\|w(k)\|^2))b$. If

$0 < \eta(a - f(\|w(k)\|^2))b) < 1$, $\left|\frac{dG}{dw(k)}\right| < 1$. Thus, $\pm\sqrt{\frac{a - f(\|w(k)\|^2)}{b}}$ are stable

points for $0 < \eta(a + f(\|w(k)\|^2))b) < 1$. The proof is completed.

Figure 7.5 illustrates the convergence of (7.2).

7.4   Chaotic Behavior

The invariant sets S_1 and S_2 guarantee the nondivergence of algorithms (7.1) and (7.2). In S_1, the algorithm (7.1) will converge to 0 if $0 < \eta(a + f(w^2)) < 2$. In invariant set S_2, the algorithm (7.2) will converge to 0 if $-2 < \eta(a - f(w^2)) < 0$ and converge to $\pm\sqrt{\frac{a - f(\|w(k)\|^2)}{b}}$ if $0 < \eta(a - f(w^2)) < 1$. Then, if $2 \leq \eta(a + f(w^2)) \leq 4$, what does the dynamical behavior of (7.1) look like? And if $1 \leq \eta(a - f(w^2)) \leq 2$, what does the dynamical behavior of (7.2) look like? In this section, we will draw the bifurcation diagram of the algorithms and calculate Lyapunov exponents of the algorithms to illustrate the chaotic behavior of the algorithms.

The bifurcation diagrams of algorithms (7.1) and (7.2) are shown in Figures 7.6 and 7.7, respectively. The bifurcation diagrams clearly show the route to chaos. Figure 7.6 shows that $\eta(a + f(w^2)) = 2$ is a bifurcation point of

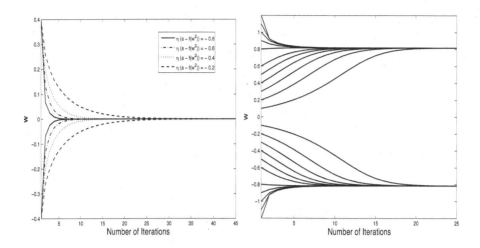

FIGURE 7.5

The algorithm (7.2) converges to 0 at the initial values $w(0) = 0.4$ or $w(0) = -0.4$ with the different $\eta(a - f(w^2))$ and $b = 3$ (left). The algorithm converge to $\pm\sqrt{\frac{a-f(\|w(k)\|^2)}{b}} = \pm 0.8165$ with $b = 3, \eta = 0.4, (a - f(\|w(k)\|^2)) = 2$ at the different initial values (right).

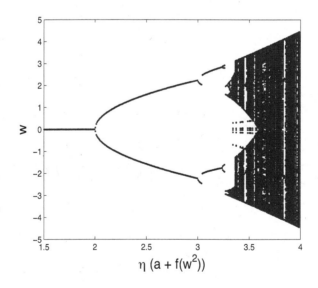

FIGURE 7.6

Bifurcation diagram of (7.1) with $w(0) = 0.01$.

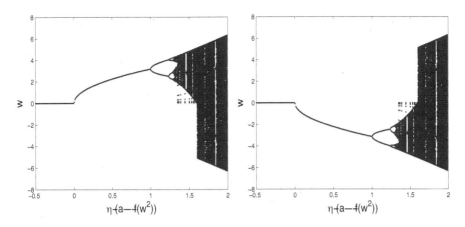

FIGURE 7.7
Bifurcation diagram of (7.2) with $w(0) = 0.01$ (left) and $w(0) = -0.01$ (right).

(7.1) where equilibria become unstable and periodic solutions show up. The bifurcation point of (7.2) is $\eta(a - f(w^2)) = 1$, as shown in Figure 7.7.

To further illustrate the chaotic behavior, the Lyapunov exponents of algorithms (7.1) and (7.2) are computed according to [2, 61]. From (7.1), let

$$y(k) = \frac{dG}{dw(k)} = 1 - \eta(a + f(w^2)) + 3\eta bw(k)^2,$$

for $0 < \eta(a + f(w^2)) < 4$ and $k \geq 0$. As for (7.2), let

$$y(k) = \frac{dG}{dw(k)} = 1 + \eta(a - f(w^2)) - 3\eta bw(k)^2,$$

for $-2 < \eta(a - f(w^2)) < 2$ and $k \geq 0$. The Lyapunov exponents can be computed by

$$\lim_{k \to +\infty} \frac{1}{k} \sum_{i=1}^{k} |y(k)|.$$

Figures 7.8 and 7.9 plot the Lyapunov exponents of algorithms (7.1) and (7.2) for each parameter value. Clearly, in Figure 7.8, as $\eta(a + f(w^2)) =\backsim 3.4$, the Lyapunov exponent becomes positive and the evolution begins to become chaotic. In Figure 7.9, as $\eta(a + f(w^2)) =\backsim 1.4$, the Lyapunov exponent becomes positive and the evolution begins to become chaotic. Thus, it is easy to see that the algorithms (7.1) and (7.2) exhibit chaotic behaviors as learning parameters increase in the invariant sets.

From the previous analysis, it is seen that the existence for chaos is because the learning parameters are chosen in certain range. Thus, it is possible to control the chaotic behavior by adjusting the learning parameters so that this algorithm can work better in practical applications.

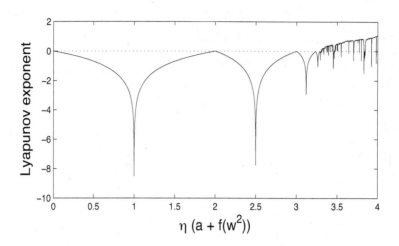

FIGURE 7.8
Lyapunov exponents of (7.1).

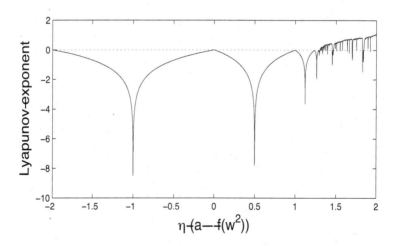

FIGURE 7.9
Lyapunov exponents of (7.2).

7.5 Conclusion

A class of ICA algorithms with constant learning rates is studied in this chapter. The invariant sets and some conditions for local stability of equilibria are derived. It is shown that the algorithms may exhibit chaotic behaviors, even diverging to infinity under some conditions. Therefore, it is required to choose proper learning parameters so that the algorithm can work more effectively.

8

Determination of the Number of Principal Directions in a Biologically Plausible PCA Model

8.1 Introduction

In the Chapters 2, 3 and 4, we have studied the convergence of PCA learning algorithms. However, to apply PCA neural networks to practical problems, an important issue to address is how to determine an appropriate number of principal directions for the PCA neural networks. The number of principal directions crucially affects the computation results of PCA neural networks. For example, in the application of PCA neural networks to image compression, if the number of principal directions is too low, the quality of the reconstructed image would be inadequate. However, if the number of principal directions is too high, an excessive computational burden would be incurred. So far, this problem, that is, how to adaptively select an appropriate number of the principal directions so as to strike an adequate balance between the quality and complexity of computations, is still largely unsolved.

In the past, the number of principal directions for PCA neural networks is usually specified based on prior knowledge. Obviously, this is not practical in many applications, especially for online ones. In this chapter, a biologically plausible PCA model is proposed to overcome this problem. The number of principal directions in this model is determined to adaptively approximate the intrinsic dimensionality of the given data set by using an improved generalized Hebbian algorithm (GHA). Thus, the given data set can be compacted to a subspace with intrinsic dimensionality through the proposed network. This allows a balance to be struck between the quality of results and the complexity of computations needed to achieve it [126].

It is known that intrinsic dimensionality of a data set reflects an appropriate number of principal directions of the data set. In more general terms, a data set $\Omega \in R^n$ is said to have an intrinsic dimensionality equal to m if its elements lie entirely within an m-dimensional subspace of R^n, where $m < n$ [26]. There are several methods to evaluate the intrinsic dimensionality of a data set, see [26, 151]. In [151], Rubio proposed an explained variance method to identify the intrinsic dimensionality m of a given data set to approach a precision α by

$$m = min \left\{ Z \in \{0, 1, \ldots, n\} \mid \sum_{p=1}^{Z} \lambda_p \geq \alpha \cdot trace(C) \right\}, \qquad (8.1)$$

where $C \in R^n$ is the covariance matrix of the given data set, $\lambda_p(i = 1, \ldots, n)$ are eigenvalues of C with $\lambda_1 \geq \lambda_2 \geq \ldots \geq \lambda_n \geq 0$. The quotient between λ_i and the trace of C is the amount of variance explained by the ith principal direction of C. Hence, α represents the amount of variance with the principal directions associated with the first m eigenvalues.

Although the intrinsic dimensionality m is just an appropriate number of principal directions for PCA neural networks, it is not easy to compute the number m in advance for online applications. In this chapter, a PCA model is proposed that is motivated by a key characteristic of biological neural networks: that excited neurons can activate neurons close to them and inhibit neurons far away from them. Using this well-known *on-center/off-surround* principle [75, 101], the proposed model together with an improved GHA algorithm is able to adapt itself. In this learning procedure, the number of active neurons, which is the same as the number of principal directions in the model, approximates the intrinsic dimensionality m of the given data set to achieve the precision α. Therefore, a good balance between the quality of computational results and the complexity of the computations of PCA neural networks can be obtained.

This chapter[4] is organized as follows. In Section 2, the biologically plausible PCA model and the improved GHA algorithm are given. Some important properties of the model are analyzed in Section 3. Simulations are carried out in Section 4. Conclusions are drawn in Section 5.

8.2 The PCA Model and Algorithm

8.2.1 The PCA Model

Suppose an input sequence

$$\{x(k) \mid x(k) \in R^n, (k = 1, 2, \ldots)\}$$

is a zero mean stationary stochastic process. Denote $C = E\left[x(k)x^T(k)\right]$, the covariance matrix of the input data set. In 1989, Sanger [156] proposed the GHA to adaptively extract the principal components from the data set. Let $\mathbf{y}(k) = W(k)x(k)$, where $W^T(k) = [\mathbf{w}_1(k), \mathbf{w}_2(k), \ldots, \mathbf{w}_{m'}(k)]$ is the weight

---

[4]Based on "Determining of the number of principal directions in a biologically plausible PCA model," by (Jian Cheng Lv, Zhang yi, and K. K. Tan), which appeared in IEEE Trans. Neural Networks, vol. 18, no. 3, pp. 910–916, May 2007. ©[2007] IEEE.

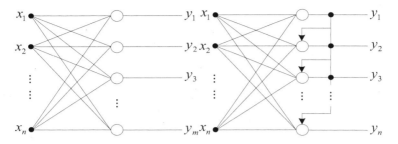

FIGURE 8.1
The original GHA model (left) and an improved GHA model with lateral connections (right).

matrix and $\mathbf{y}(k) = [y_1(k), y_2(k), \ldots, y_{m'}(k)]^T$ is the output vector. The GHA PCA model is presented in Figure 8.1 (left) and the GHA learning algorithm is as follows:

$$W(k+1) = W(k) + \eta \left[W(k)C_k - LT\{W^T(k)C_k W(k)\}W(k) \right], \tag{8.2}$$

for $\eta > 0$ and $k \geq 1$, where $C_k = x(k)x^T(k)$. The operator $LT[\cdot]$ sets all the elements above the diagonal of its matrix argument to zero. By (8.2), $W^T(k) = [\mathbf{w}_1(k), \mathbf{w}_2(k), \ldots, \mathbf{w}_{m'}(k)]$ will converge to the principal directions of the input data set. Here, $m' << n$ is the number of extracted principal directions of the given data set and the number of neurons in this model. Clearly, the number m' must be determined in advance to use the PCA neural network. This requirement is not practical in applications, especially for online ones.

In this chapter, an improved PCA model is proposed as shown in Figure 8.1 (right). In this model, the number of the extracted principal directions will be determined to adaptively approximate the intrinsic dimensionality of the given data set to reach a required precision. The neurons in this model are arranged in a line with lateral connections. The lateral connections serve as passages through which the output of neurons can be transmitted to other neurons so that the activation and inhibition can occur. There are two classes of neurons in this model: active neurons and resting neurons. All active neurons always line up in front of this model, and the first neuron is always at the center of all active ones. After the active neurons are excited by input with a certain magnitude, either the first resting neuron will be activated or the last active neuron will be inhibited, based on the *on-center/off-surround* biological perspective [75, 101]. The number of active neurons, which is just the number of the principal directions for this PCA model, approximates the intrinsic dimensionality of the data set via the biological procedure.

With the online observation C_k (1.10), clearly, $\lim_{k \to \infty} C_k = C$ with $C_0 = \mathbf{0}$. The improved model can be presented mathematically as $\mathbf{y}(k) =$

$W(k)x(k), (k \geq 1)$, where $W^T(k) = [\mathbf{w}_1(k), \mathbf{w}_2(k), \ldots, \mathbf{w}_n(k)]$ is the weight matrix and $\mathbf{y}(k) = [y_1(k), y_2(k), \ldots, y_n(k)]^T$ is the output vector. To update the net weights, an improved GHA learning algorithm is given by

$$W(k+1) = W(k) + \eta \Pi(k) [W(k)C_k - LT\{W^T(k)C_k W(k)\}W(k)], \tag{8.3}$$

for $\eta > 0$ and $k \geq 1$, where $\Pi(k)$ is a diagonal matrix. $\Pi(k)$ serves as a controller at time instance k, which determines whether a neuron is active or resting. The value of the diagonal element of $\Pi(k)$ is always 1 or 0 and 1 always appear along the first column. If the value of the diagonal element is 1, the corresponding neuron is active. Otherwise, the neuron is resting. The initial values of all elements of $\Pi(1)$ are 0, except $\Pi_{11}(1) = 1$, that is, only the first neuron is active at the initial time. It is just $\Pi(k)$, which determines the number of the extracted principal directions for the improved PCA model at time instance k.

The key difference between the original GHA (8.2) and the improved GHA (8.3) is that $\Pi(k)$ is added in the improved GHA. By changing the value of diagonal elements of $\Pi(k)$, the algorithm (8.3) will adaptively converge to the first m principal directions of the input data set, that is, $\tilde{W}^T = [\mathbf{w}_1, \mathbf{w}_2, \ldots, \mathbf{w}_m]$, so that the dimensionality of the data set can be reduced to the m-dimensionality online with precision α. Here, an important problem to address is how to change the value of diagonal elements of $\Pi(k)$ so that the number of principal directions for the PCA model approximates the intrinsic dimensionality of the input data set in an online manner. The following subsection will discuss this problem.

8.2.2   Algorithm Implementation

The value of diagonal elements of $\Pi(k)$ is adaptively adjusted with time for the number of the extracted principal directions to approximate the intrinsic dimensionality of the input data set. The changes to the value of diagonal elements of $\Pi(k)$ effectively implement the *on-center/off-surround* procedure. In this section, the implementation of the algorithm will be described.

First, define two functions as follows:

$$\sigma(x) = \begin{cases} 1, & \text{if } x \geq 0 \\ 0, & \text{otherwise} \end{cases} \qquad \bar{\sigma}(x) = \begin{cases} 1, & \text{if } x \geq 0 \\ -1, & \text{otherwise.} \end{cases}$$

Denote $l(k)$ as the number of active neurons at time instance k, which is just the number of the extracted principal directions for the proposed PCA model. $l(k)$ is also the number of 1 in diagonal elements of $\Pi(k)$, with $l(1) = 1$.

In Chapter 4, it has been proved that $\mathbf{w}_i(k)$ converges to the ith principal direction of C_k with $k \to \infty$, and $y_i^2(k) = \mathbf{w}_i^T(k)C_k\mathbf{w}_i(k)$ converges to λ_i. Thus, $\sum_{i=1}^{l(k)} y_i^2(k)/trace(C_k)$ approximates the amount of variance expressed by the first $l(k)$ principal directions of C_k. Suppose the number $l(k)$ of active neurons approximates the intrinsic dimensionality of a given data set as $k \to \infty$. $f_1(k)$ represents the threshold of intrinsic dimensionality of the data set at time instance k according to Rubio's method [151]. At time instance k, if the amount of variance expressed by the first $l(k)$ principal directions of C_k doest not reach the required precision, that is, $\alpha \cdot trace(C_k) > \sum_{i=1}^{l(k)} y_i^2(k)$, then $f_1(k) = 1$. In this case, the first resting neuron, which is close to active neurons, may be activated. Otherwise, $f_1(k) = -1$ in which case the last active neuron may be inhibited.

$$f_1(k) = \bar{\sigma} \left[\alpha \cdot trace(C_k) - \sum_{i=1}^{l(k)} y_i^2(k) \right], \qquad (8.4)$$

for $k \geq 1$, where α is the approximated precision and $trace(C_k)$ denotes the trace of matrix C_k.

Clearly, the number of active neurons cannot increase indefinitely and cannot also decrease to 0. Here, $f_2(k)$ serves as the threshold of neighborhood. Let $n - l(k)$ be the maximum length of *neighborhood* to which the neuron may be activated. The threshold of the neighborhood is defined as

$$f_2(k) = \sigma(\sigma(n - l(k) - 1) - f_1(k)), \qquad (8.5)$$

for all $k \geq 1$. If and only if $f_2(k) = 1$, a neuron may be activated. Clearly, $f_1(k)$ and $f_2(k)$ will jointly serve as the switch of the neighborhood.

If only $f_1(k)$ is used as an activation or inhibition condition, the number $l(k)$ of active neurons will oscillate around the intrinsic dimensionality of the input data set. This will be illustrated in Section 5. To overcome this problem, the $f_3(k)$ is defined as follows:

$$f_3(k) = \sigma \left\{ \sigma \left[\sum_{i=1}^{l(k)-1} y_i^2(k) - \alpha \cdot trace(C_k) \right] + f_2(k) \right\}, \qquad (8.6)$$

for all $k \geq 1$. From (8.4) and (8.6), clearly, if $\sum_{i=1}^{l(k)} y_i^2(k) > \alpha \cdot trace(C_k)$ and $\sum_{i=1}^{l(k)-1} y_i^2(k) \leq \alpha \cdot trace(C_k)$, then $f_3(k) = 0$ and $f_1(k) = -1$. In this case, activation or inhibition will not take place anymore. It just serves to fulfill a stop criterion. If $\sum_{i=1}^{l(k)-1} y_i^2(k) > \alpha \cdot trace(C_k)$, then $f_3(k) = 1$ and $f_1(k) = -1$ in which case the last active neuron may be inhibited at the time instance k.

Next, the following rules reflect the *on-center/off-surround* principle.

$$\Pi(k) = \begin{bmatrix} 1 & 0 & \cdots & \overbrace{0}^{l(k)} & \cdots & 0 \\ 0 & 1 & \cdots & 0 & \cdots & 0 \\ \vdots & \vdots & \vdots & \vdots & \vdots & \vdots \\ 0 & 0 & \cdots & 1 & \cdots & 0 \\ 0 & 0 & \cdots & 0 & \cdots & 0 \\ \vdots & \vdots & \vdots & \vdots & \vdots & \vdots \\ 0 & 0 & \cdots & 0 & \cdots & 0 \end{bmatrix},$$

$$\begin{aligned} A. \quad & l(k+1) = l(k) + f_1(k) \cdot f_2(k) \cdot f_3(k), & (8.7) \\ B. \quad & i = l(k) + \sigma\left(l(k+1) - l(k) - 1\right), & (8.8) \\ C. \quad & \Pi_{ii}(k+1) = \Pi_{ii}(k) + (l(k+1) - l(k)). & (8.9) \end{aligned}$$

Equation (8.7) computes the number of active neurons at time instance $k+1$ and (8.8) determines the diagonal element of $\Pi(k)$, which will be updated. The diagonal elements of $\Pi(k)$ are updated in (8.9).

8.3 Properties

The key difference between the original GHA (8.2) and the improved GHA (8.3) is that the matrix $\Pi(k)$ is included in the improved GHA. Clearly, the algorithm (8.3) does not change the convergence property of (8.2). In Chapter 4, the convergence of (8.2) has been studied. In this chapter, we will not discuss this problem.

However, by (8.3), (8.7), (8.8), and (8.9), the number of principal directions for the proposed PCA model does not need to be specified in advance as in the original GHA (8.2). The number can be adaptively determined to approximate the intrinsic dimensionality of the given data set. This feature is especially useful for online applications. Some properties of this algorithm will be presented, which can guarantee the validity.

In this section, some important properties of (8.3) are given as follows:

Theorem 8.1 *If $l(1) = 1$, then $1 \leq l(k) \leq n, (k \geq 1)$.*

Proof: From (8.4), (8.5), and (8.6), the values of $f_1(k), f_2(k)$, and $f_3(k)$ are, respectively,

$$f_1(k) = 1 \text{ or } -1; \quad f_2(k) = 1 \text{ or } 0; \quad f_3(k) = 1 \text{ or } 0,$$

for all $k \geq 1$. Clearly, $f_1(k) \cdot f_2(k) \cdot f_3(k) = 1$ or 0 or -1. From (8.7), we have

$$|l(k+1) - l(k)| \leq 1, \text{ for } k \geq 1. \tag{8.10}$$

From (8.4), if $l(k) = n$, $f_2(k) = \sigma(0 - f_1(k))$. From (8.5), it follows that $f_2(k) = 0$, if $f_1(k) = 1$ and $f_2(k) = 1$, if $f_1(k) = -1$. Thus, $f_1(k) \cdot f_2(k) \cdot f_3(k) = 0$ or -1. Clearly, from (8.7) and (8.10), it holds that $l(k+1) \leq l(k) \leq n, (k \geq 1)$.

If $l(k) = 1$ and $l(k) < n$, there are two cases to be considered.

Case 1: If $\alpha \cdot trace(C_k) > \sum_{i=1}^{l(k)} y_i^2(k)$, from (8.4), (8.5), it follows that $f_1(k) = 1, f_2(k) = 1, f_3(k) = 1$. Thus, $f_1(k) \cdot f_2(k) \cdot f_3(k) = 0$ or 1. From (8.7), $l(k+1) \geq l(k) \geq 1$.

Case 2: If $\alpha \cdot trace(C_k) < \sum_{i=1}^{l(k)} y_i^2(k)$, then $f_1(k) = -1$ and $f_2(k) = 1$. Since

$$\sum_{i=1}^{l(k)-1} y_i^2(k) - \alpha \cdot trace(C_k) = -\alpha \cdot trace(C_k) < 0, \quad (l(k) = 1),$$

from (8.6), $f_3(k) = 0$. So, $f_1(k) \cdot f_2(k) \cdot f_3(k) = 0$. From (8.7) and (8.10), it holds that $l(k+1) = l(k) \geq 1, (k \geq 1)$. The proof is completed.

Theorem 8.2 *If $l(1) = 1$, $\Pi_{ii}(1) = 1(i = 1)$, and $\Pi_{ii}(1) = 0(i = 2, \ldots, n)$, then it holds that (1) the diagonal elements of $\Pi(k)$ is always 1 or 0; (2) 1 always appears in the first columns in $\Pi(k)$ at time instance k; (3) $l(k)$ equals the number of 1 in $\Pi(k)$ at time instance k.*

Proof : From (8.4), (8.5), and (8.6), the values of $f_1(k), f_2(k)$, and $f_3(k)$ are, respectively,

$$f_1(k) = 1 \text{ or } -1; \quad f_2(k) = 1 \text{ or } 0; \quad f_3(k) = 1 \text{ or } 0,$$

for all $k \geq 1$. Clearly, $f_1(k) \cdot f_2(k) \cdot f_3(k) = 1$ or 0 or -1. From (8.7), it follows that

$$l(k+1) - l(k) = 1 \text{ or } 0 \text{ or } -1.$$

There are three cases to be discussed.

Case 1: If $l(k+1) - l(k) = 1$ at time instance k, then $f_1(k) = f_2(k) = f_3(k) = 1$. From (8.8), it follows for $k \geq 1$ that

$$i = l(k) + 1 = l(k+1) \text{ and } \Pi_{ii}(k+1) = \Pi_{ii}(k) + 1. \tag{8.11}$$

Case 2: If $l(k+1) - l(k) = -1$ at time instance k, then $f_2(k) = f_3(k) = 1$ and $f_1(k) = -1$. From (8.8), it follows for $k \geq 1$ that

$$i = l(k) = l(k+1) + 1 \text{ and } \Pi_{ii}(k+1) = \Pi_{ii}(k) - 1. \tag{8.12}$$

Case 3: If $l(k+1) - l(k) = 0$ at time instance k, then any one of $f_1(k), f_2(k), f_3(k)$ is 0. From (8.8), it follows for $k \geq 1$ that

$$i = l(k) \text{ and } \Pi_{ii}(k+1) = \Pi_{ii}(k) + 0. \tag{8.13}$$

From (8.11), (8.12), and (8.13), it is not difficult to reach the conclusion

$$\begin{cases} \Pi_{ii}(k) = 0, & \text{if } i = l(k+1) = l(k) + 1, \\ \Pi_{ii}(k) = 1, & \text{if } i = l(k) = l(k+1) + 1, \\ \Pi_{ii}(k) = 1, & \text{if } i = l(k) = l(k+1), \end{cases}$$

for $k \geq 1$. So, it holds that

$$\begin{cases} \Pi_{ii}(k+1) = 1, & \text{if } i = l(k+1) = l(k) + 1, \\ \Pi_{ii}(k+1) = 0, & \text{if } i = l(k) = l(k+1) + 1, \\ \Pi_{ii}(k+1) = 1, & \text{if } i = l(k) = l(k+1), \end{cases} \qquad (8.14)$$

for $k \geq 1$. From (8.14), it is easy to arrive at the theorem. The proof is completed.

Theorems 8.1 and 8.2 guarantee the validity of (8.3). $l(k)$, the number of active neurons, always vary over a certain range. Since 1 in $\Pi(k)$ always appears in the first column in $\Pi(k)$ at time instance k, the first principal directions of the input data set can be obtained.

Theorem 8.3 *The number $l(k)$ of principal directions for the improved PCA model approximates the intrinsic dimensionality of data set C as $k \to \infty$ with the precision α.*

Proof. According to the convergence results of GHA algorithm in chapter 4, from Theorem 8.2, we have

$$\lim_{k \to \infty} \sum_{i=1}^{l(k)} y_i^2(k) = \lim_{k \to \infty} \sum_{i=1}^{l(k)} w_i(k) C_k w_i^T(k) = \sum_{i=1}^{l(\infty)} \lambda_i,$$

where λ_i is eigenvalue of covariance matrix C. From (8.4), (8.5), and (8.7), if

$$\alpha \cdot trace(C) > \sum_{i=1}^{l(k)} \lambda_i \text{ and } l(k) < n, \text{ as } k \to \infty,$$

it follows that $l(k+1) = l(k) + 1$, if $f_1(k) = 1$. From (8.4), (8.5), (8.6), and (8.7), if

$$\sum_{i=1}^{l(k)-1} \lambda_i > \alpha \cdot trace(C) \text{ and } l(k) > 1, \text{ as } k \to \infty,$$

it follows that $l(k+1) = l(k) - 1$, if $f_1(k) = 1$. Clearly, it holds that

$$\alpha \cdot trace(C) < \sum_{i=1}^{l(\infty)} \lambda_i(k) \text{ and } \alpha \cdot trace(C) > \sum_{i=1}^{l(\infty)-1} \lambda_i(k). \qquad (8.15)$$

Thus, from Theorem 8.2 and Rubio's explained variance method, (8.15) shows that the number $l(k)$ of principal directions for the improved PCA model will approximate the intrinsic dimensionality of the data set C with precision α [151]. The proof is completed.

Theorem 8.3 shows that the number of the extracted principal directions for the improved PCA model approximates the intrinsic dimensionality m of C as $k \to \infty$ with the precision α. Thus, by (8.3), the given data set can be compacted into a m-dimensional subspace with the precision α.

8.4   Simulations

In this section, some examples will be provided to illustrate the properties of (8.3). For the algorithm (8.3), the learning rate η affects the learning speed and convergence of the algorithm. Since the algorithm (8.3) does not change the convergence of the original GHA, which has been discussed in Chapter 4, in this section, we will only illustrate that the number of principal directions for our PCA model thus determined will adaptively approximate the intrinsic dimensionality of a given data set with $\eta = 0.1$ so that the data set can be compacted into a subspace with precision α.

8.4.1   Example 1

The first example will illustrate intrinsic dimensionality and the typical problem encountered by the traditional PCA algorithms. Consider an off-line case. The covariance matrix of an off-line data set is generated randomly as

$$C = \begin{bmatrix} 0.1712 & 0.1538 & 0.0976 & 0.0367 & 0.0796 & 0.1290 \\ 0.1538 & 0.1386 & 0.0873 & 0.0330 & 0.0726 & 0.1164 \\ 0.0976 & 0.0873 & 0.0675 & 0.0325 & 0.0436 & 0.0708 \\ 0.0367 & 0.0330 & 0.0325 & 0.0198 & 0.0168 & 0.0257 \\ 0.0796 & 0.0726 & 0.0436 & 0.0168 & 0.4110 & 0.0621 \\ 0.1290 & 0.1164 & 0.0708 & 0.0257 & 0.0621 & 0.9860 \end{bmatrix}.$$

It is easy to obtain its eigenvalues as $1.0534, 0.4621, 0.2579, 0.0200, 0.0060, 0.0000$. Its trace is equal to 1.9741. By (8.1), the intrinsic dimensionality m can be calculated. If $\alpha = 0.84$, then $m = 2$, and $m = 3$ if $\alpha = 0.98$. Let $C_k = C$ for all $k \geq 1$. By (8.3), Figure 8.2 (left) illustrates that the number of the extracted principal directions converges to the intrinsic dimensionality 3 of C with precision $\alpha = 0.98$. Thus, by using the proposed PCA model, the data set can be compressed to three-dimensional space with a precision of $\alpha = 0.98$. In addition, if the function $f_3(k)$ is not used, the number of principal directions will oscillate around the intrinsic dimensionality of 3. The result is shown in Figure 8.2 (right).

Clearly, suppose that we use the traditional PCA algorithm to reduce the

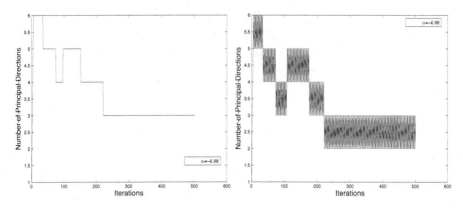

FIGURE 8.2
The number of principal directions approximates the intrinsic dimensionality with $\alpha = 0.98$ (left) and the number will oscillate around the intrinsic dimensionality if $f_3(k)$ is not used (right).

dimensionality of the data set to reach a certain precision, the dimensionality m must be specified in advance. In off-line cases, it may be possible to compute m. However, this requirement is not practical and easily met in online applications. The improved PCA model (8.3) can adaptively reduce the dimensionality of a data set to any precision without the need to specify the dimensionality m in advance.

8.4.2   Example 2

Several gray scale images will be used in online simulations for illustration of the algorithm and its advantages over the traditional one. First, the 512×512 pixel image for Lenna, in Figure 1.5, is decomposed into 4096 blocks that do not intersect one another. With the input data set (1.9) and online observation C_k (1.10), the algorithm (8.3) is used to compact the original picture.

To illustrate the effectiveness of (8.3), by (8.1), we calculate the intrinsic dimensionality of the given data set $\{x(k)\}$ using MATLAB 7.0. The intrinsic dimensionality is obtained as $m = 3$ if $\alpha = 0.95$ and $m = 8$ if $\alpha = 0.98$. Then, the improved algorithm (8.3) is used to compute the principal directions of the given data set. Figure 8.3 (upper) shows that the number of principal directions converges to 3 with $\alpha = 0.95$, that is, $\tilde{W}^T = [\mathbf{w}_1, \mathbf{w}_2, \mathbf{w}_3]$. The number of principal directions converges to 8 with $\alpha = 0.98$ in Figure 8.3 (bottom), that is, $\tilde{W}^T = [\mathbf{w}_1, \mathbf{w}_2, \dots, \mathbf{w}_8]$. In addition, if the function $f_3(k)$ is not used, the number $l(k)$ of principal directions oscillates around the intrinsic dimensionality of the given data set. The result is shown in Figure 8.3 (middle).

Clearly, after the algorithm (8.3) has converged, the improved PCA model, $\mathbf{y}(k) = \tilde{W}x(k), (k \geq 1)$, can compact the original data set into an eight-

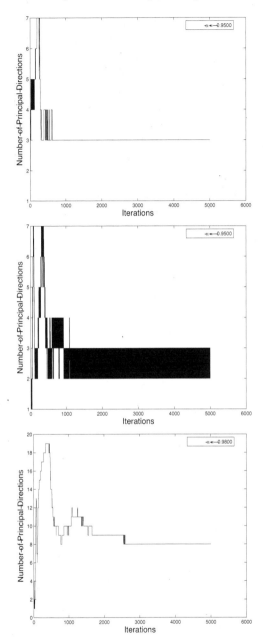

FIGURE 8.3
The number of principal directions approximates the intrinsic dimensionality
with $\alpha = 0.95$ (upper); the number will oscillate around the intrinsic dimensionality if $f_3(k)$ is not used (middle), and the number of principal directions
approximates the intrinsic dimensionality with $\alpha = 0.98$ (bottom).

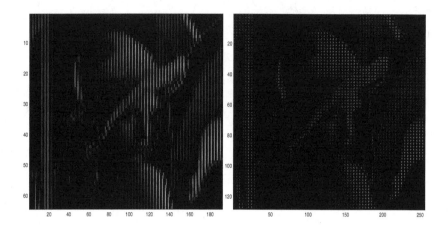

FIGURE 8.4
The Lenna image is compacted into an three-dimensional subspace with $\alpha = 0.95$ (left), and the image is compacted into an eight-dimensional subspace $\alpha = 0.98$ (right).

FIGURE 8.5
The reconstructed image with $\alpha = 0.95$ (left), and the reconstructed image with $\alpha = 0.98$ (right).

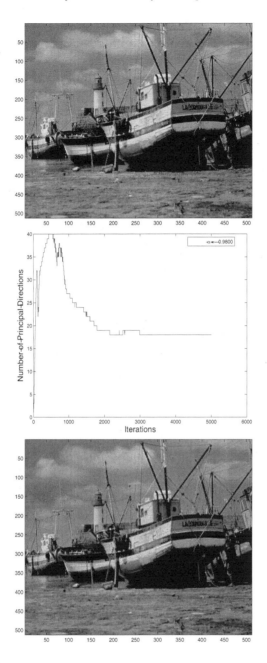

FIGURE 8.6

The original image for *Boat* (upper); the number of principal directions converging to the intrinsic dimensionality 18 with $\alpha = 0.98$ (middle), and the reconstructed image (bottom).

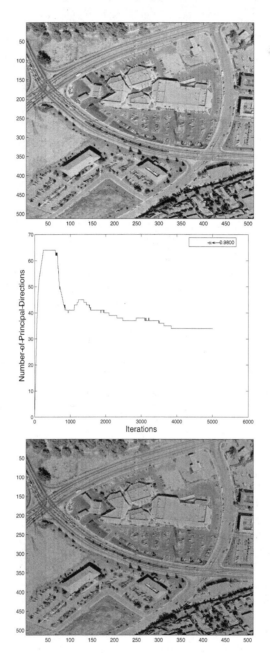

FIGURE 8.7

The original image for *Aerial* (upper); the number of principal directions converging to the intrinsic dimensionality 35 with $\alpha = 0.98$ (middle), and the reconstructed image (bottom).

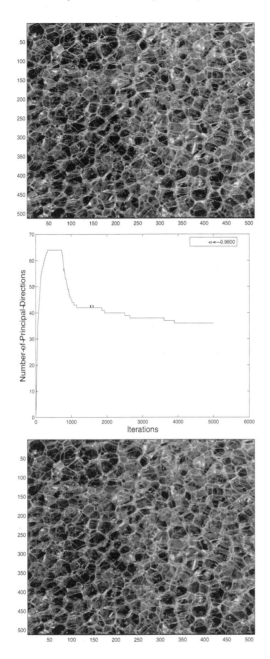

FIGURE 8.8
The original image for *Plastic bubbles* (upper); the number of principal directions converging to the intrinsic dimensionality 37 with $\alpha = 0.98$ (middle), and the reconstructed image (bottom).

dimensional subspace with the precision $\alpha = 0.98$. When $\alpha = 0.95$, the original data set can also be compacted to a three-dimensional subspace. Figure 8.4 shows the compacted result of the original image. In Figure 8.4 (left), each of the small rectangles (1×3 pixels) in the compacted image represents a compacted three-dimensional vector from the 64-dimensional vector. The compacted image in Figure 8.4 (left) is represented by a three-dimensional vector set with 4096 vectors, that is, $Y = \{y(k) \in R^3, k = 1, 2, \ldots, 4096\}$. In Figure 8.4 (right), each of the small rectangles (2×4 pixels) in the compacted image represents a compacted eight-dimensional vector from the 64-dimensional vector. The compacted image in Figure 8.4 (right) is represented by an eight-dimensional vector set with 4096 vectors, that is, $Y = \{y(k) \in R^8, k = 1, 2, \ldots, 4096\}$. Clearly, the original image is represented by a 64-dimensional vector set with 4096 vectors. This shows the great advantage of the improved GHA method.

The reconstructed images can be simply obtained by using the reconstructing equation [77]:

$$\tilde{X} = \tilde{W}^T Y,$$

where $\tilde{W}^T$ is the trained weight matrix of the improved GHA, Y is the vector set of the compressed image, and $\tilde{X}$ represents the reconstructed 64-dimensional vector set. Figure 8.5 shows the reconstructed image with $\alpha = 0.95$ (left) and $\alpha = 0.98$ (right).

Next, the 512×512 gray images for *Boat, Aerial, Plasticbubbles* from USC SIPI Image Database, will be used to further illustrate the proposed algorithm (8.3). The original images are shown in the upper pictures in Figure 8.6, 8.7, and 8.8. The middle pictures show that the number of principal directions converges to the intrinsic dimensionality of the given data set with the precision $\alpha = 0.98$. The reconstructed images are shown in the bottom pictures.

8.5   Conclusion

In this chapter, a PCA model with improved GHA algorithm is proposed. In this model, an appropriate number of extracted principal directions can be determined to adaptively approximate the intrinsic dimensionality of the input data set. The model is suitable for online applications to reduce the dimensionality of a given data set to the intrinsic dimensionality of the data set with precision α. Since an appropriate number of principal directions will strike a balance between the quality of the results and complexity of computation of PCA neural networks, the algorithm is especially useful for online applications. Simulations have been conducted to verify the theoretical results.

9

Multi-Block-Based MCA for Nonlinear Surface Fitting

9.1 Introduction

In [133], clustered blockwise PCA takes advantage of the spatiotemporal correlation and localized frequency variations that are typically found in data sets. Therefore, the algorithm not only achieves greater efficiency in the resulting representation of the visual data, but also successfully scales PCA to handle large data sets.

Since MCA can also be applied to any given data set by breaking the data into pieces and vectorizing them as PCA does, taking advantage of MCA neural network and localized frequency variations in data, we propose a method of multi-block-based MCA neural network for nonlinear surface fitting. MCA is first applied to blocks individually, which are obtained by breaking the given data set into pieces and vectorizing them, treating each block as if they were independent and identically distributed. Then the reconstructed results of the blocks are grouped into a new data set, and MCA is applied again to the grouped data set to obtain the final result. We call this method multi-block-based MCA. According to research on natural image statistics [20], for almost any visual data, different frequency variations in the data tend to be spatially localized in regions that are distributed over the data. Therefore, this multi-block-based MCA would achieve better accuracy, compared with applying MCA directly on the data set (called MCA direct method).

In this chapter, we use an improved Oja+'s MCA neural network rather than the traditional matrix algebraic approaches to track the minor component (MC). As the MCA neural network is able to extract MC from input data adaptively without calculating the covariance matrix in advance, the computational complexity is simplified compared to that of matrix algebraic approaches; hence, it is suitable for dealing with online data. At the same time, the Oja+'s MCA algorithm is improved by using an adaptive learning rate. This learning rate converges to a nonzero constant so that the evolution will become faster with time. However, the adaptive learning rate can prevent the weight from deviating from the right evolution direction by adjusting itself. Thus, the global convergence could be guaranteed so that the selection of the initial parameters is simple.

This chapter is organized as follows. The traditional MCA method is reviewed, and the improved Oja+'s MCA neural network is presented in Section 2. In Section 3, we propose a method of multi-block-based MCA neural network. Some simulation results will be given to illustrate the effectiveness and accuracy of the proposed method in Section 4. Finally, in Section 5, the conclusion follows.

9.2 MCA Method

9.2.1 Matrix Algebraic Approaches

Let $X = \{x(k)|x(k) \in R^n (k = 0, 1, 2, \ldots)\}$ be an n-dimensional zero-mean, Gaussian stationary data set with the nonnegative definite covariance matrix $C = E\{x(k)x^T(k)\}$. Let λ_i and $w_i (i = 1, \ldots, n)$ denote the eigenvalues and the corresponding eigenvectors of C, respectively. We shall arrange the orthonormal eigenvectors such that the corresponding eigenvalues are in a nondecreasing order: $0 < \lambda_1 \le \lambda_2 \le \ldots \le \lambda_n$, then $w_i^T X$ is the ith smallest minor component, where w_i is the corresponding eigenvector to λ_i. To obtain the eigenvectors and eigenvalues, we can either apply eigen decomposition to the covariance matrix of the data matrix or directly apply SVD to the data matrix.

Applying SVD, we can get

$$X = UDV^T$$

in which, U's columns are eigenvectors of XX^T, and V's columns are eigenvectors of $X^T X$, while D is a diagonal matrix with the diagonal elements d_i, the singular values, which are the square roots of the eigenvalues of $X^T X$.

9.2.2 Improved Oja+'s MCA Neural Network

In [135], Oja proposed to use a linear neural network in Figure 1.5 for computing the minor component of the covariance matrix associated with a given stochastic data set. The Oja+'s MCA learning algorithm can be described by the following stochastic difference equations:

$$\begin{aligned}
w(k+1) &= w(k) - \eta(k)[C_k w(k) \\
&\quad -w^T(k)C_k w(k)w(k) \\
&\quad +\mu(\|w(k)\|^2 - 1)w(k)],
\end{aligned} \tag{9.1}$$

where μ is a constant and $\eta(k) > 0$, and $C_k = x(k)x^T(k)$ is an online observation of the covariance matrix defined by $C = E\left[x(k)x^T(k)\right]$. To afford the tracking capability of the adaptive algorithm, the online observation C_k (1.10) is used.

If learning rate $\eta(k)$ converges to zero, this is not practical in applications as discussions in [40, 195, 202, 210]. Furthermore, Oja+ algorithm may suffer from the sudden divergence [48]. In this chapter, we propose to use the following adaptive learning rate:

$$\eta(k) = \frac{\xi}{w^T(k)(\mu I - C_k)w(k)}, \quad (0 < \xi < 1),$$

for $k \geq 0$. The proposed algorithm is given as follows:

$$
\begin{aligned}
w(k+1) \;=\; & w(k) - \frac{\xi}{w^T(k)(\mu I - C_k)w(k)}[C_k w(k) \\
& -w^T(k)C_k w(k)w(k) \\
& +\mu(\| \, w(k) \, \|^2 - 1)w(k)],
\end{aligned}
\tag{9.2}
$$

for $k \geq 0$, where $0 < \xi < 1$. The condition $\mu > \lambda_n$ is required for the stability reason. Clearly, this learning rate converges to a non-zero constant so that the evolution will become faster with time. However, the adaptive learning rate can prevent the weight from deviating from the right evolution direction by adjusting itself. Thus, the global convergence could be guaranteed [125]. After convergence, $w^T(k+1)x(k)$ is the extracted minor component. Considering all the advantages stated previously, we adopt the improved Oja+'s algorithm in multi-block-based MCA.

9.2.3 MCA Neural Network for Nonlinear Surface Fitting

In the case of nonlinear surface fitting, many nonlinear curves and hypersurfaces models can be expressed as [185]

$$
\begin{aligned}
a_1 f_1(\mathbf{x}) + a_2 f_2(\mathbf{x}) + \ldots + a_m f_m(\mathbf{x}) + c_0 = 0, \\
\mathbf{x} = [x_1, x_2, \ldots, x_n]^T,
\end{aligned}
$$

where $\mathbf{x}$ is the original coordinate points and $f_i(\mathbf{x})$ are some basis functions, that is, polynomials of the elements of $\mathbf{x}$. In the quadratic case

$$a_1 x_1^2 + a_2 x_1 x_2 + a_3 x_2^2 + a_4 x_1 + a_5 x_2 + c_0 = 0,$$

function $f_i(\mathbf{x})$ are defined as follows:

$$
\begin{aligned}
f_1(\mathbf{x}) = x_1^2, \quad f_2(\mathbf{x}) = x_1 x_2, \quad f_3(\mathbf{x}) = x_2^2, \\
f_4(\mathbf{x}) = x_1, \quad f_5(\mathbf{x}) = x_2.
\end{aligned}
$$

When Gaussian noise is added to the observation data, we can use MCA neural network to estimate the right parameters $[a_1, a_2, \ldots, a_n]^T$ and c_0. Let $\mathbf{f} = [f_1(\mathbf{x}), f_2(\mathbf{x}), \ldots, f_m(\mathbf{x})]^T$. We train the neural network by learning algorithm (9.2), with $x(t)$ being replaced by $\mathbf{f}' = \mathbf{f} - E(\mathbf{f})$.

Let us consider the case when the observation data are a three-dimensional data set, which comes from an ellipsoid. Generate a noise observation data set $D = \{(x_i, y_i, z_i), i = 0, 1, \ldots, N\}$ by adding Gaussian noise to the original data points. We can have a parameterized model

$$a_1 x^2 + a_2 y^2 + a_3 z^2 + c_0 = 0 \tag{9.3}$$

to be fitted to the data points by estimating the parameters a_1, a_2, a_3. When applying the improved Oja+s MCA neural network, the steps to get the fitted surface are as follows:

1. Calculate the means $e_f = [e_1, e_2, e_3]^T, e_1 = E(x'), e_2 = E(y'), e_3 = E(z')$.

2. Subtract the means from each data point, that is, $\xi_1^{(i)} = x'_i - E(x'), \xi_2^{(i)} = y'_i - E(y'), \xi_3^{(i)} = z'_i - E(z')$, to get a zero-mean data set.

3. Initial the weight vector $w(0) = [w_1, w_2, w_3]^T$ by random elements, and let $x(j) = [\xi_1^{(j)}, \xi_2^{(j)}, \xi_3^{(j)}]^T$, j be a random integer with equal probability being any integer of $\{1, 2, \ldots, N\}$.

4. Use (9.2) to adaptively adjust w, so that $[a_1, a_2, a_3]^T = w$ and $c_0 = -E^T(\mathbf{f})\mathbf{x}$, where a_1, a_2, a_3, c_0 are the corresponding parameters in (9.3).

5. Renormalize the parameters to the case of nonzero means for the data points by substituting a_1, a_2, a_3, c_0 to (9.3), that is, $a_1 = -w_1/c_0, a_2 = -w_2/c_0, a_3 = -w_3/c_0, c_0 = -\hat{w}^T e_f$.

9.3   Multi-Block-Based MCA

In MCA direct method, MCA is applied directly to the whole data set once to obtain the smallest weight vector [185]. For multi-block-based MCA on surface fitting, we implement the fitting process in the following steps:

1. Partition. To divide the original data set into smaller blocks, then apply the improved Oja+'s MCA neural network to each block individually, and subsequently reconstruct the blocks using converged weight vectors, respectively.

2. Grouping. To group all the reconstructed blocks into one block according to the order in which they were before partition.

3. Fitting. To apply the improved Oja+'s MCA neural network to the grouped block.

To explain the specific process, we still take the three-dimensional observation data set for example. Given a data set $D = \{(x_i, y_i, z_i), i = 0, 1, \ldots, N\}$, which is generated by adding Gaussian noise to a data set that comes from an ellipsoid, where $N = W \times H$, for W and H are the numbers of distinct values along x- and y- axes, respectively. We divide the data points into $N/(B_x \times B_y)$ small blocks in terms of x-y plane, so that each block size is $B_x \times B_y$. Then

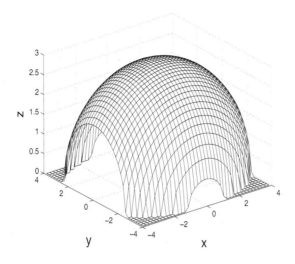

FIGURE 9.1
The original data set taken from the ellipsoid segment (9.4).

each block turns into a $3 \times (B_x \times B_y)$ matrix. MCA learning algorithm is subsequently applied to the blocks individually to obtain weight vectors, as was specified in Section 9.2.3. We use these weight vectors to fit the parameterized model (9.3) to the data points in each block, respectively. After that, we group all the reconstructed blocks into one block to form a $3 \times (W \times H)$ matrix, in the same order as they are partitioned before. Finally, MCA learning algorithm is applied again to the grouped data set, that is, the $3 \times (W \times H)$ to extract the smallest minor component; hence, the fitting is completed by fitting (9.3) to the grouped data set.

9.4 Simulation Results

9.4.1 Simulation 1: Ellipsoid

In this simulation, the original data set $D = \{(x_i, y_i, z_i), i = 0, 1, \ldots, 1600\}$ is sampled from an ellipsoid segment

$$0.04x^2 + 0.0625y^2 + 0.1111z^2 = 1,$$
$$(-4 < x \le 4, -4 < y \le 4, 0 < z). \tag{9.4}$$

Considering that in computer vision problems the influence of noise on variables x, y can usually be neglected, while the z variable has been contami-

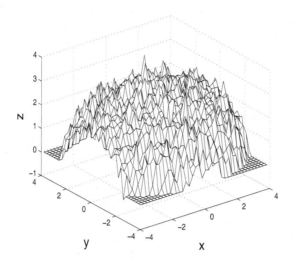

FIGURE 9.2
The noise-disturbed data set used for surface fitting, which is obtained by adding Gaussian noise with zero means and variance $\sigma^2 = 0.3$ in the way given by (9.5).

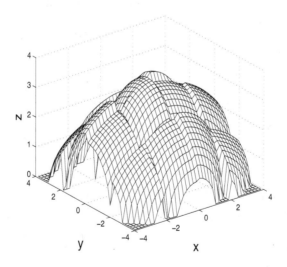

FIGURE 9.3
The data points obtained after applying the improved Oja+'s algorithm to each 10×10 block with $\mu = 40$.

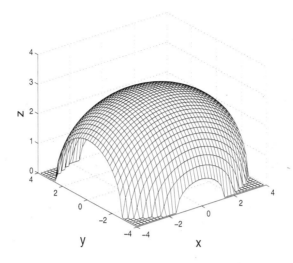

FIGURE 9.4
The data points obtained after applying the improved Oja+'s algorithm to the grouped data with $\mu = 40$, that is, the final result of surface fitting by using multi-block-based MCA on the data set given by Figure 9.2.

nated by noise much stronger than x, y, we generate a noise observation data set $D' = \{(x'_i, y'_i, z'_i), i = 0, 1, \ldots, 1600\}$ by adding Gaussian noise to z-axis observation data as follows:

$$z'_i = z_i + n_z, \quad x'_i = x_i, \quad y'_i = y_i, \quad i = 1, 2, \ldots, 1600. \tag{9.5}$$

The Gaussian noise is with zero means and variance $\sigma^2 = 0.3$. Figures 9.1 and 9.2 show the original ellipsoid segment and the noise-disturbed data set D', respectively.

We partition the noise-disturbed data points into 16 blocks, and each block size is 10×10. MCA is then applied to each block by using the improved Oja+'s algorithm (9.2) according to the process mentioned in Section 9.2.3. After reconstructing the data points in each block, we group all the blocks into one to form a matrix. The result after applying the improved Oja+'s MCA algorithm to blocks individually is shown in Figure 9.3. Subsequently, we apply the improved Oja+'s algorithm (9.2) one more time to the grouped matrix. Figure 9.4 shows the final result of surface fitting by using multi-block-based MCA.

Comparing with applying MCA direct method, the simulation results of multi-block-based MCA on the same data are shown in Table 9.1. The absolute error between the true parameter vector w and the estimated $\hat{w}$, defined as $\varepsilon_w = \|\hat{w} - w\|$, is used to compare the accuracy.

TABLE 9.1

The solution obtained by different methods on the same noisy data set. The average values of w obtained from 20 simulations are given.

	w_1	w_2	w_3	ε_w
True values	0.0400	0.0625	0.1111	
MCA direct method	0.0368	0.0590	0.1125	0.0051
Multi-block-based MCA	0.0375	0.0604	0.1084	0.0045

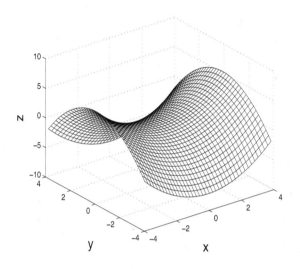

FIGURE 9.5

The original data points taken from the saddle segment (9.6).

9.4.2 Simulation 2: Saddle

To illustrate the effectiveness of our method, we give another surface fitting simulation on a data set $D = \{(x_i, y_i, z_i), i = 0, 1, \ldots, 1600\}$ sampled from a saddle

$$z = 0.3615x^2 - 0.4820y^2,$$
$$(-4 < x \le 4, \quad -4 < y \le 4). \tag{9.6}$$

this time the noise-disturbed data set $D' = \{(x_i', y_i', z_i'), i = 0, 1, \ldots, 1600\}$ this time is obtained by adding Gaussian noise with zero means and variance $\sigma^2 = 0.5$ to z-axis observation data, in the way given by (9.5). The original saddle and the noise-disturbed data set D' are, respectively, shown in Figures 9.5 and 9.6.

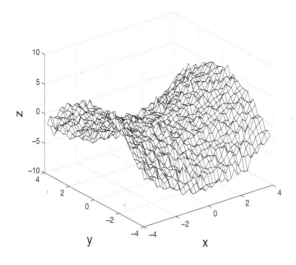

FIGURE 9.6
The noise-disturbed data set used for surface fitting, which is obtained by adding Gaussian noise with zero means and variance $\sigma^2 = 0.5$ in the way given by (9.5).

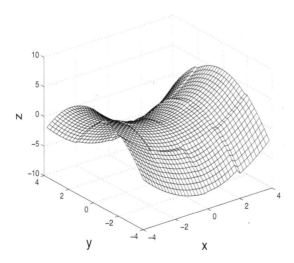

FIGURE 9.7
The data points obtained after applying the improved Oja+'s algorithm to each 10×10 block with $\mu = 30$.

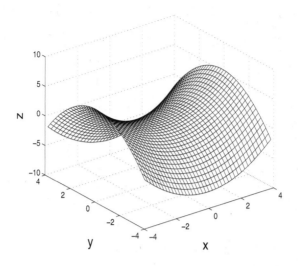

FIGURE 9.8
The data points obtained after applying the improved Oja+'s algorithm to
the grouped data with $\mu = 50$, that is, the final result of surface fitting by
using multi-block-based MCA on the data set given by Figure 9.6.

Different from simulation 1, we generally fit a parameterized model

$$a_1 x62 + a_2 y^2 + a_3 z + c_0 = 0$$

to the noise-disturbed data points, that is, $\mathbf{f} = \mathbf{f} - E(\mathbf{f})$ is used to train the
neural network, where $\mathbf{f} = [x^2, y^2, z]$. In this simulation, as the original real
data set has zero means, the true c_0 is zero, while the estimated c_0 we obtain
is close to zero. The block size we use in this simulation is 10×10. Figure 9.7
shows the result after partition and applying MCA to each block, and Figure
9.8 shows the final result of fitting using multi-block-based MCA. Table 9.2
shows the simulation results of applying multiblock-based MCA and MCA
direct method, respectively, on the same saddle data set disturbed by strong
Gaussian noise.

From the simulation results (Tables 9.1 and 9.2), we see that the estimated
solution by multi-block-based MCA is considerably good, and the multi-block-
based MCA outperforms the MCA direct method in terms of accuracy.

TABLE 9.2

The solution obtained by different methods on the same noisy data set. The average values of w obtained from 20 simulations are given.

	w_1	w_2	w_3	ε_w
True values	0.3615	0.4820	1.0000	
MCA direct method	0.3646	0.4844	1.0000	0.0046
Multi-block-based MCA	0.3644	0.4846	1.0000	0.0044

9.5 Conclusion

In this chapter, a multi-block-based MCA method is proposed for nonlinear surface fitting. The proposed method shows good performance in terms of accuracy, which is confirmed by the simulation results.

10

A ICA Algorithm for Extracting Fetal Electrocardiogram

10.1 Introduction

Fetal electrocardiogram (FECG) extraction is an interesting and challenging problem in biomedical engineering. The antepartum FECG is usually obtained through skin electrodes attached to the mother's body [89]. Unfortunately, the desired fetal heartbeat signals appear at the electrode output buried in an additive mixture of undesired disturbances, such as the mother's heartbeat signal with extremely high amplitude, the noise caused by mother's respiration, and electronic equipment. Appropriate signal processing techniques are required to recover the wanted FECG components from the corrupted potential recordings.

Many approaches have been reported to get the FECG, such as singular value decomposition(SVD) [95], multireference adaptive noise cancellation [200], and so on. But the results are not satisfactory or highly sensitive to electrode placement. Recently, it was shown in [107] that the problem can be modeled as blind source separation (BSS) [43, 83]. It was reported that the results by BSS were more satisfactory than those by classical methods [5, 174, 200]. However, separating all the sources from a large number of observed sensor signals takes a long time and is not necessary, so the blind source extraction (BSE) method is preferred. The basic objective of BSE is estimating one or part of source signals from their linear mixtures. Such signals are usually with specified stochastic properties and bring the most useful information.

Many BSE algorithms use the property of sparseness [209] or high-order statistics [46, 115, 205] to extract a specific signal. High-order methods address the BSE problem in a completely blind context, since they require few assumptions aside from the statistical independence of the sources. The second-order statistics (SOS) [11, 45, 206], however, operate in a semiblind setting, since their derivation usually requires the certain additional assumptions made on the nature of the source signals, such as statistical nonstationarity of the sources, presence of time correlations in stationary signals, or cyclostationarity. Nonetheless, such information is available in practical applications and should be exploited. Furthermore, the high-order approaches often have higher computational load compared with the second-order methods. Thus, the versatile extraction algorithms based on second-order statistics become popular.

Barros et al. [11] presented an objective function by using blind signal extraction along with prior information about the autocorrelation property of the FECG from the noisy-free measurements. We extend the work for the noisy case. Theoretical analysis of the new algorithm is also given in this chapter [111].

This chapter is organized as follows. In Section 2, problem formulation is given. A new algorithm is proposed and theoretical analysis is given in Section 3. Section 4 presents the simulations. Conclusions are drawn in Section 5.

10.2   Problem Formulation

Let us denote $s(k) = [s_1(k), \ldots, s_n(k)]^T$ as the source signals vector and by $x(k) = [x_1(k), \ldots, x_n(k)]^T$ as the observed signals vector. Thus, the mixture can be written as

$$x(k) = As(k) + n(k), \tag{10.1}$$

where A is an $n \times n$ nonsingular matrix and $n(k)$ is additive noise. In general, the noise $n(k)$ is assumed to be uncorrelated with the source signals and its covariance matrix is given by

$$E\{n(k)n^T(k - \triangle k)\} = \begin{cases} 0, & for \ \triangle k \neq 0 \\ R_n, & for \ \triangle k = 0 \end{cases}. \tag{10.2}$$

Normally, we further assume that $R_n = \sigma_n^2 I$, where $\sigma_n^2 I$ is the variance of the noise. There are several approaches to estimate σ_n^2. If the number of mixtures is larger than the number of sources, we can use a subspace method to estimate σ_n^2, which in this case represents the smallest eigenvalue of $E\{x_k x_k^T\}$ [76], or we can use an adaptive principal component analysis method to real-time estimation of σ_n^2 [84].

We assume that the source signals have temporal structure and different autocorrelation functions, but they do not necessarily have to be statistically independent. Denote by s_i the fetal heartbeat signal to extract, which satisfies the following relation for the specific time delay τ_i :

$$\begin{cases} E\{s_i(k)s_i(k - \tau_i)\} \neq 0 \\ E\{s_i(k)s_j(k - \tau_i)\} = 0 \quad \forall j \neq i \end{cases}. \tag{10.3}$$

Let $y(k) = w^T x(k)$, where w is the weight vector to be estimated such that $y(k)$ recovers the specific source signal s_i. Define the error function as in [11]:

$$\varepsilon(k) = y(k) - by(k - \tau_i), \tag{10.4}$$

where b is a parameter to be determined and τ_i can be obtained in advance. A simple solution is to calculate the autocorrelation $E[x_j(t)x_j(t - \tau_i]$ of sensor signals as a function of the time delay and find the one corresponding to the pick of $E[x_j(t)x_j(t - \tau_i]$ [11]. For simplicity, denote $y_k \triangleq y(k)$, $y_{\tau_i} \triangleq y(k - \tau_i)$, $x_k \triangleq x(k)$, $x_{\tau_i} \triangleq x(k - \tau_i)$.

The aim of extracting the desired FECG is then converted to seek appropriate parameters w and b to minimize the objective function:

$$f(w, b) = E\{\varepsilon^2(k)\} - (1 + b^2)\sigma_n^2. \tag{10.5}$$

10.3   The Proposed Algorithm

Calculating the derivative of the objective function (10.5) with respect to w and b, we have the following updating rule:

$$\begin{cases} w(k+1) = w(k) - \eta 2E\left\{(x_k - bx_{\tau_i})(x_k - bx_{\tau_i})^T\right\}w(k) \\ b(k+1) = b(k) - \eta[w^T(k)E\left\{-x_kx_{\tau_i}^T - x_{\tau_i}x_k^T + 2bx_{\tau_i}x_{\tau_i}^T\right\}w(k) - 2b\sigma_n^2] \end{cases} \tag{10.6}$$

followed by a normalization of the Euclidean norm of vector $w(k)$ to unit to avoid the trivial solution $w(k) = 0$. In Eq. (10.6), η is a small positive learning rate. Moreover, we can assume without losing generality that the sensor data are prewhitened; thus, $E\{x_kx_k^T\} = I$. The algorithm updates b together with w to make the objective function (10.5) decrease and eventually extract the desired signals.

Theorem 10.1 *The objective function (10.5) decreases along the iteration rule (10.6) with sufficiently small η.*

Proof: Denote $D_k \triangleq E\left\{(x_k - bx_{\tau_i})(x_k - bx_{\tau_i})^T\right\}$, we have

$$f[w(k), b(k)] = w^T(k)D_kw(k) - (1 + b^2)\sigma_n^2. \tag{10.7}$$

Then with the notation $f_{w(k)} \triangleq 2E\left\{(x_k - bx_{\tau_i})(x_k - bx_{\tau_i})^T\right\}w(k)$ and $f_{b(k)} \triangleq w^T(k)E\left\{-x_kx_{\tau_i}^T - x_{\tau_i}x_k^T + 2bx_{\tau_i}x_{\tau_i}^T\right\}w(k) - 2b\sigma_n^2$, we have

$$\begin{aligned} f[w(k+1), b(k+1)] &= E\left[w(k) - \eta f_{w(k)}\right]^T E\left\{\left[x_k - \left(b(k) - \eta f_{b(k)}\right)x_{\tau_i}\right]\right. \\ &\quad \times \left[x_k - \left(b(k) - \eta f_{b(k)}\right)x_{\tau_i}\right]^T\right\}\left[w(k) - \eta f_{w(k)}\right] \\ &\quad - (1 + (b - \eta f_{b(k)})^2\sigma_n^2) \\ &= f[w(k), b(k)] - 4\eta w^T(k)D_k^2w(k) \\ &\quad - \eta f_{b(k)}^2 + o(\eta), \end{aligned} \tag{10.8}$$

where $-f^2_{b(k)} \leq 0$ for all k. Since the matrix D^2_k is positive semidefinite in practice, we have $w^T(k)D^2_k w(k) \geq 0$. Thus, the objective declines. The proof is completed.

Theorem 10.2 *Denote a performance vector by $c = A^T w_*$, where w_* is the weight vector estimated by the updating rule (10.6). If for that specific time delay τ_i, relations (10.3) hold, then $c = \beta e_i$, where β is a nonzero scalar and e_i is such a vector that the ith element is ± 1 and the other elements are 0.*

Proof: The w^* and b^* estimated by the algorithm satisfy

$$w_* = E\left\{ (x_k - b_* x_{\tau_i})(x_k - b_* x_{\tau_i})^T \right\} w_*, \tag{10.9}$$

and

$$w^T_* E\left\{ x_k x^T_{\tau_i} + x_{\tau_i} x^T_k - 2b_* x_{\tau_i} x^T_{\tau_i} \right\} w_* - 2b\sigma^2_n = 0. \tag{10.10}$$

From (10.10), we have

$$c^2_i E\left\{ s_i(k)s_i(k - \tau_i) \right\} - b_* \sum_{j=1}^{N} c^2_j E\left\{ s^2_j(k - \tau_i) \right\} = 0.$$

If $c_i = 0$, then $c_j = 0, \forall j \neq i$, which is impossible. Assume that $c_i \neq 0$, multiplying both sides of equation (10.9) by nonsingular matrix A^T, a basic condition for the resolvability of the BSE problem, it gives that

$$A^T w_* = A^T E\left\{ x_k x^T_k - b_* \left(x_k x^T_{\tau_i} + x_{\tau_i} x^T_k \right) + b^2_* x_{\tau_i} x^T_{\tau_i} \right\} w_*.$$

Hence, taking into account, that $E\{x_k x^T_k\} = I$, we have

$$c_j = \left[1 + \left(A^T A \right)_j b^2_* E\left\{ s^2_j(k - \tau_i) \right\} + b^2_* \sigma^2_n \right] c_j, \quad \forall j \neq i,$$

where $\left(A^T A \right)_j$ denotes the jth diagonal element. Note that $(A^T A)_j E\{s^2_j(k - \tau_i)\} + \sigma^2_n = 1$; we know it is clear that $c_j = 0, \forall j \neq i$. Taking $c_i \neq 0$ into consideration, we have $c = \beta e_i$. Theorem 10.2 indicates that the algorithm (5) actually leads to the extraction of the desired source signals.

10.4 Simulation

To show the validity of the proposed algorithm, we use the well-known ECG data set [52]. The ECG signals were recorded over 10 seconds from eight cutaneous electrodes located in different points of a pregnant woman's body (Figure 10.1). The sampling frequency was 250 Hz. One can see that the fetus's heartbeat is weaker and faster compared with that of the mother.

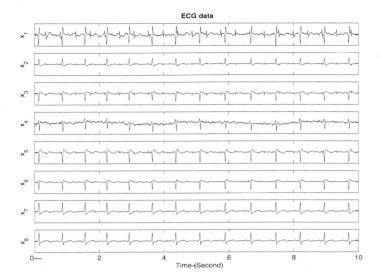

FIGURE 10.1

ECG signals taken from a pregnant woman. x_1-x_5 and x_6-x_8 correspond to abdominal and thoracic measurements from a pregnant woman, respectively.

It was important to estimate the appropriate τ_i. It could be carried out by examining the autocorrelation of the observed signals. In Figure 10.2, this function is shown for the case of the first observed signal x_1, where the fetal influence was clearly stronger than in the other signals. The autocorrelation of x_1 shown in Figure 10.2 presents some peaks. Since we do not know which is the most appropriate, it is important to have a previous idea of the most probable value of τ_i. One can make use of the fact that a fetal heart rate is around 120 beats per second. A heart rate of 120 beats per second means that the heart should strike every 0.5 second. By carefully examining the autocorrelation of the sensor signal x_1, we found that it had a peak at 0.448 second, corresponding to $\tau_i = 112$ [11].

The variance σ_n^2 of the additive noise $n(k)$ is estimated to be 0.3. Setting the learning rate $\eta = 0.1$ we run the algorithm (10.6) and the Barros's algorithm in [11]. The recovering result is shown in Figure 10.3. Obviously, the extracted FECG by our algorithm is clearer, while the one by Barros's algorithm is mixed by noise. Furthermore, we can see from Figure 10.4 that the value of the objective (10.5) monotonically declines with the iteration.

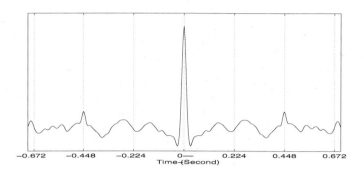

FIGURE 10.2
Autocorrelation of signal of x_1 of Figure 10.1.

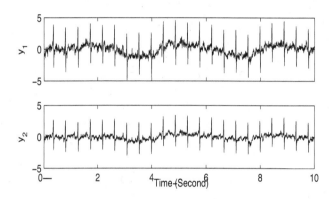

FIGURE 10.3
Extracted FECG: y_1 is extracted by Barros's algorithm; y_2 is by our algorithm
(10.6).

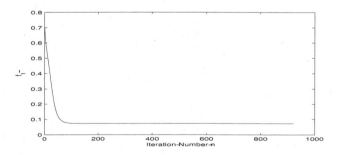

FIGURE 10.4
The value of the objective function (10.5). It declines with the parameters of
w and b updated by the new algorithm.

10.5   Conclusion

Extracting the FECG signal from the composite maternal ECG signal is of great interest. This chapter propose a new algorithm for FECG extraction with additive noise. The updating rule makes the objective function monotonically decrease and extracts the desired signals when the objective function gets minima. The theoretical analysis and simulation results verify the validity of the proposed algorithm.

11

Rigid Medical Image Registration Using PCA Neural Network

11.1 Introduction

It is common for patients to undergo multiple tomographic radiological imaging for the purpose of medical diagnosis. These multi-modality images provide complementary information. However, it is difficult for doctors to fuse these images exactly because of the variations in patient orientation. For this reason, there has been considerable interest in using image registration techniques to transfer all the information into a common coordinate frame. The objective of this study is to find a transformation that best aligns the float image (unregistered image) with the reference image.

Many registration methods have been developed for decade years [177]. In general, these methods are based either on original images or on feature images. Among these methods, commonly used registration techniques include Fourier transform [62], mutual information [69], and cross-correlation method [10]. Recently, the registration method using mutual information is very popular. This method can register multimodality medical images accurately without preprocessing. However, the mutual information computation is complex and costs much time. So there has been considerable interest in finding new methods to rapidly register medical images.

The proposed method based on feature images accomplishes registration by simply aligning feature images' first principal directions and centroids [160]. The objective of this method is to find parameters (rotation angle, translations along the x- and y- axis) for registration. This method consists of three steps:

(1) *Feature extraction.* For computed tomography-magnetic resonance (CT-MR) registration, extracted contours are used as feature images. However, for MR-MR registration, threshold segmented images are used as feature images. The next two steps are based on feature images obtained in this step.

(2) *Computing the rotation angle.* The proposed method uses a principal component analysis (PCA) neural network to compute the first principal direction of the reference feature image and that of the float

feature image, then the angle between the two directions is simply calculated, which is used as the rotation angle.

(3) *Computing translation.* Translations are calculated as subtracting float feature image's centroid from reference feature image's centroid.

In the procedure above, the crucial step is to calculate the first principal directions by using PCA neural network [125, 134, 136, 172, 195]. Since a PCA neural network can compute the principal direction more easily, the computation is simplified. Moreover, the principal direction usually converges to a unit vector, which makes the computation of rotation angle simpler.

11.2　Method

The proposed method consists of three steps: (1) feature extraction; (2) Computing the rotation angle; and (3)computing translations. The two-dimensional images in Figures 11.1 and 11.2 are used to illustrate the procedure.

11.2.1　Feature Extraction

In general, the implementation of medical image registration is either on original images or on feature images. The proposed registration method is implemented on feature images.

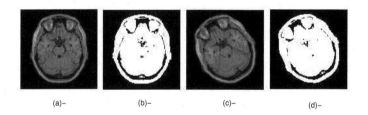

(a)–　　　　　(b)–　　　　　(c)–　　　　　(d)–

FIGURE 11.1
Threshold segmentation: (a) Reference image and (b) its reference feature image with threshold 40. (c) Float image and (d) its float feature image with threshold 40.

For MR-MR images registration, threshold segmented images are used as feature images. The procedure of threshold segmentation consists of two steps: (1) setting a threshold θ to lie within $[0, 255]$; (2) comparing image's intensity with θ, where values larger than θ are set to be 1, and all the others are set to be 0. Figure 11.1 shows the threshold segmentation results. Figure 11.1(a) and Figure 11.1(c) are MR images of a patient taken at different times. Figure 11.1(a) is used as reference image, and Figure 11.1(c) is the corresponding float image. Figure 11.1(b) and Figure 11.1(d) are feature images of Figure 11.1(a) and Figure 11.1(c) with threshold 40, respectively. In this chapter, Figure 11.1(b) is called reference feature image and Figure 11.1(d) is called float feature image.

For CT-MR registration, contours are used as feature images. In this chapter, contour tracking method is applied to extracting contour information. The detailed procedure of this method can be found in [182]. Figure 11.2 shows contour extraction results with contour tracking method. Figure 11.2(a) is a patient's CT image, which is used as the reference image. The MR image in Figure 11.2(c) is used as the float image. Figure 11.2(b) and 11.2(d) are feature images of Figure 11.2(a) and 11.2(c), respectively.

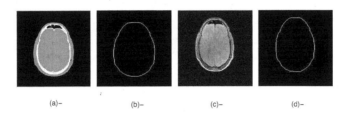

(a)–   (b)–   (c)–   (d)–

FIGURE 11.2
Contour extraction: (a) CT image (reference image) and (b) its reference feature image. (c) MR image (float image) and (d) its float feature image.

11.2.2   Computing the Rotation Angle

Computing the rotation angle is an important step for rigid medical image registration. In this section, we present the way of finding the rotation angle that relates to feature images obtained in the previous section.

If an image is rotated, it is clear that the rotation angle is equal to the rotation angle of its first principal direction. The mathematic fundament is given as follows:

First, we give a brief review of the traditional PCA [8, 165]. Let vector $X_i = [x_i, y_i]^T$ denote points of an image. Clearly, its first principal direction W is defined as follows:

$$W = \arg\max_{V} \left(V^T S V \right), \tag{11.1}$$

where S is the covariance matrix and is given by

$$S = \frac{1}{N} \sum_{i=1}^{N} (X_i - \mu)(X_i - \mu)^T, \mu = \frac{1}{N} \sum_{i=1}^{N} X_i.$$

Clearly, the principal direction W is the eigenvector associated with the largest eigenvalue of the covariance matrix S.

Suppose an image is rotated by angle θ. Let vector $\tilde{X}_i = [\tilde{x}_i, \tilde{y}_i]^T$ denote points of the rotated image. Clearly, $\tilde{X}_i$ can be represented as follows:

$$\tilde{X}_i = R \cdot X_i, \tag{11.2}$$

where R is the transform matrix and is given by

$$R = \begin{bmatrix} \cos(\theta) & -\sin(\theta) \\ \sin(\theta) & \cos(\theta) \end{bmatrix}.$$

Then, the first principal component $\tilde{W}$ of the rotated image is computed as follows:

$$\tilde{W} = \arg\max_{\tilde{V}} \left\{ \tilde{V}^T \mathrm{E} \left[\left(\tilde{X} - \tilde{\mu} \right) \left(\tilde{X} - \tilde{\mu} \right)^T \right] \tilde{V} \right\},$$

By (11.2), it follows that

$$
\begin{aligned}
\tilde{W} &= \arg\max_{\tilde{V}} \left\{ \left(R^T \tilde{V} \right)^T \mathrm{E} \left[(X - \mu)(X - \mu)^T \right] \left(R^T \tilde{V} \right) \right\} \\
&= \arg\max_{\tilde{V}} \left\{ \left(R^T \tilde{V} \right)^T S \left(R^T \tilde{V} \right) \right\}. \tag{11.3}
\end{aligned}
$$

From (11.1) and (11.3), it is easy to get that

$$\tilde{W} = RW. \tag{11.4}$$

Equations (11.2) and (11.4) imply that the rotated image and its first principal component have undergone the same rotation transform.

In our method, the first principal directions of the reference feature image and float feature image are computed by using PCA neural network, then it is easy to get the rotation angle, which is the angle between the two first principal directions.

Registration methods using neural networks have been developed in recent years [63, 132, 146]. In the proposed method, Oja's PCA neural network is used to compute the first principal direction.

Oja's network could compute the principal direction of the covariance matrix S associated with a given stochastic data set [134]. Suppose the input sequence $\{x(k)|x(k) \in R^n (k = 0, 1, 2, \ldots)\}$ is a zero mean stationary stochastic process and let $C_k = x(k)x^T(k)$, then the Oja's PCA learning algorithm can be described by the following stochastic difference equations:

$$w(k + 1) = w(k) + \eta[C_k w(k) - w^T(k)C_k w(k)w(k)], \tag{11.5}$$

for η is the learning rate and $w(k) = [w_1(k), w_2(k), \ldots, w_n(k)]^T$. If learning rate η satisfies some simple condition, the w will converge to the first principal direction [125, 195]. The single layer neural model shown in Figure 1.5 is used to extract the first principal direction, and (11.5) is its learning algorithm.

In our applications, the reference feature image and float feature image are quantified into the input sequence $\{r(k)\}$ and $\{f(k)\}$, respectively, where

$$r(k) = \begin{pmatrix} x(k) \\ y(k) \end{pmatrix} \text{ or } f(k) = \begin{pmatrix} x(k) \\ y(k) \end{pmatrix},$$

for all $k \geq 0$. By using Oja's algorithm (11.5), it is easy to obtain reference image's first principal direction $W_R = [r_x, r_y]^T$ with $\|W_R\| = 1$ and float feature image's first principal direction The angle between the two vectors is calculated as follows:

$$Ab_\theta = \left| \arccos \left(W_R^T \cdot W_F \right) \right|, \tag{11.6}$$

where Ab_θ is the absolute value of rotation angle.

Since rotation angle is usual within $[-30°, 30°]$, the sign of rotation angle can be obtained according to

$$\theta = \begin{cases} \text{sgn}\,(f_y/f_x) \cdot Ab_\theta, & \text{sgn}\,(r_y/r_x) \neq \text{sgn}\,(f_y/f_x) \\ \text{sgn}\,(r_y/r_x - f_y/f_x) \cdot Ab_\theta, & \text{sgn}\,(r_y/r_x) = \text{sgn}\,(f_y/f_x) \end{cases}, \tag{11.7}$$

where function sgn is defined as following

$$\text{sgn}(x) = \begin{cases} 1, & x > 0 \\ 0, & x = 0 \\ -1, & x < 0 \end{cases}.$$

11.2.3   Computing Translations

After deriving rotation angle, we present the way of computing translations of registration.

Consider a data set $\left\{X_i | X_i = [x_i, y_i]^T, 1 \leq i \leq N\right\}$. With the knowledge of physics, its centroid $[X_c, Y_c]$ is computed by the following:

$$X_c = \frac{1}{N} \sum_i x_i, Y_c = \frac{1}{N} \sum_i y_i. \tag{11.8}$$

From (11.8), float feature image's centroid $\left[X_c^f, Y_c^f\right]$ and reference feature image's centroid $[X_c^r, Y_c^r]$ be obtained. Translations are calculated as subtracting float feature image's centroid from the reference feature image's centroid:

$$delX = X_c^f - X_c^r, delY = Y_c^f - Y_c^r,$$

where $delX$ and $delY$ are translations along x-axis and y-axis, respectively.

11.3 Simulations

This section includes two examples to illustrate the proposed method.

11.3.1 MR-MR Registration

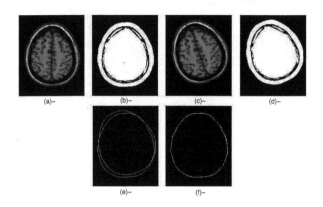

FIGURE 11.3
MR-MR image registration: (a) Reference image and (b) its feature image with threshold 31. (c) Float image and (d) its feature image with threshold 31. (e) Comparison of contours before registration versus (f) after registration

Figure 11.3 shows the result of MR-MR registration. Figure 11.3(a) is used as the reference image, and Figure 11.3(c) is the corresponding float image, which is obtained by a sequential rotation and translation operation with the MRIcro software (http://www.psychology.nottingham.ac.uk). Figure 11.3(b) and 11.3(d) are feature images of Figure 11.3 (a) and 11.3(c), respectively. With PCA neural network, the first principal direction of Figure 11.3(b) is $[-0.037, -0.999]^T$ and that of Figure 11.3(d) is $[-0.376, -0.926]^T$. Then, by (11.6) and (11.7), the rotation angle is $-20.11°$. By subtracting the centroid of Figure 11.3(d) from the centroid of Figure 11.3(b), the translations along the x- and y-axis are -1.1 and 8.1 pixels, respectively.

To facilitate the examination of the proposed method, registered image's contour and reference image's contour are compared in Figure 11.3(f). Clearly, the two contours almost overlap, which illustrates our method is promising.

11.3.2 CT-MR Registration

Images provided in the framework of "The Retrospective Registration Evaluation Project" (http://www.vuse.vanderbilt.edu/image/registration/) are used in this section. Figure 11.4(a) shows the result of CT-MR registration. Figure

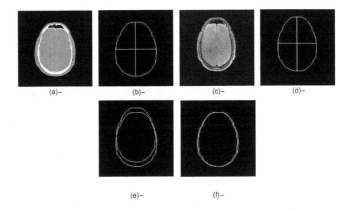

FIGURE 11.4
CT-MR images registration: (a) Patient's head CT image and (b) its contour, centroid, principal components. (c) Patient's head MR image and (d) its contour, centroid, principal components. (e) Comparison of contours before registration versus (f) after registration

11.4(a) is the reference image and 11.4(c) is the corresponding float image. Figure 11.4(b) and 11.4(d) are feature images of Figure 11.4(a) and 11.4(c), respectively. The reference image's contour and registered image's contour are compared in Figure 11.4(f). Clearly, the two contour almost overlap, which illustrates the proposed method is promising for CT-MR image registration.

11.4 Conclusion

This section proposes an automatic method for CT-MR registration or MR-MR registration. The proposed method based on feature images accomplishes registration by simply aligning the first principal directions and centroids of feature images. Since a PCA neural network is used to compute the first principal directions of feature images, the registration is simple and efficient.

Bibliography

[1] H. M. Abbas and M. M. Fahmy. Neural model for Karhunen-Loeve transform with application to adaptive image compression. *IEE Proceedings-I*, 140(2):135–143, April 1993.

[2] H. N. AGIZA. On the analysis of stability, bifurcation, chaos and chaos control of Kopel map. *Chaos, Solitons & Fractals*, 10:1909–1916, 1999.

[3] T. Akuzawa. Nested newton's method for ICA and post factor analysis. *IEEE Trans. Signal Processing*, 51(3):839–852, Mar. 2003.

[4] S. Amari, T. Chen, and A. Cichocki. Stability analsysis of learning algorithms for blind source separation. *Neural Networks*, 10(8):1345–1351, 1997.

[5] K. V. K. Ananthanag and J. S. Sahambi. Investigation of blind source separation methods for extraction of fetal ECG. In *IEEE CCECE 2003. Canadian Conference on Electrical and Computer Engineering, Montreal, Canada*, pages 2021–2024, 2003.

[6] B. R. Andrievskii and A. L. Fradkov. Control of chaos: methods and application. I. methods. *Automation and Remote Control*, 51(5):673–713, May 2003.

[7] S. Bannour and R. M. Azimi-Sadjadi. Principal component extraction using recursive least squares learning. *IEEE Trans. Neural Networks*, 6(2):457–469, March 1995.

[8] P. Bao and H. Hung. PCA neural network for JPEG image enhancement. *Signals, System, and Computer*, 2:976–980, 1999.

[9] S. Barbarossa, E. Daddio, and G. Galati. Comparision of optimum and linear prediction technique for clutter cancellation. In *Proc. Inst. Elect. Eng.*, pages 134:277–282, 1987.

[10] D. I. Barnea and H. F. Silverman. A class of algorithms for fast digital image registration. *IEEE Trans. Computer*, 21:179–186, 1996.

[11] A. K. Barros and A. Cichocki. Extraction of specific signals with temporal structure. *Neural Computation*, 13(9):1995–2003, Sep. 2001.

[12] S. Bartelmaos and K. Abed-Meraim. Fast minor component extraction using Givens rotations. *Electronics Letters*, 43(18):1001–1003, Aug. 2007.

[13] A. Bell and T. J. Sejnowski. Edges are the independent components of natural scenes. In *In Advances in Neural Information Processing 9 (NIPS 96)*, pages 831–837. MIT Press, 1997.

[14] A. Benveniste, P. Priouret, and M. Métivier. *Adaptive Aglorithms and Stochastic Approximations*. Springer, 1990.

[15] D. Bertrand, D. Guibert, J. P. Melcion, and G. Guc. Evaluation of the transition matrix for comminuting pea seeds in an impact mill using a linear neural network. *Powder Technology*, 105:119–124, 1999.

[16] L. Billings, E. M. Bollt, and I. B. Schwartz. On the unification of line processes, outlier rejection, and robust statistics with applications in early vision. *Physical Review Letters*, 88:234101(1–4), Nov. 2002.

[17] C. M. Bishop. *Pattern Recognition and Machine Learning*. Springer, New York, 2006.

[18] M. J. Black and A. Jepson. Eigentracking: robust matching and tracking of objectings using view-based representation. In *Proceedings, Volume I. Lecture Notes in Computer Science 1064, ECCV, Cambridge, UK*, pages 329–342. Springer, 1996.

[19] M. J. Black and A. Rangarajan. On the unification of line processes, outlier rejection, and robust statistics with applications in early vision. *International Journal of Computer Vision*, 19(1):57–91, 1996.

[20] D. Blanco and B. Mulgrew. ICA in signals with multiplicative noise. *IEEE Trans. Signal Processing*, 53(8):2648–2657, Aug. 2005.

[21] D. Blanco, D. P. Ruiz, E. A. Hernández, and M. C. Carrión. A fourth-order stationarity and ergodicity conditions for harmonic processes. *IEEE Trans. Signal Processing*, 52(6):1641–1649, June 2004.

[22] T. Blaschke and L. Wiskott. CuBICA: independent component analysis by simultaneous third- and fourth-order cumulant diagonalization. *IEEE Trans. Signal Processing*, 52(5):1250–1256, May 2004.

[23] T. Blaschke, T. Zito, and L. Wiskott. Independent slow feature analysis and nonlinear blind source separation. *Neural Computation*, 19:994–1021, 2007.

[24] J. F. C. L. Cadima and I. T. Jolliffe. Variable selection and the interpretation of principal subspaces. *Journal of Agricultural, Biological, and Environmental Statistics*, 6(1):62–79, Mar. 2001.

[25] E. R. Caianiello, M. Marinaro, and S. RamPone. Outline of a linear neural network. *Neurocomputing*, 12:187–201, 1996.

[26] F. Camastra and A. Vinciarelli. Estimating the intrinsic dimension of data with a fractal-based method. *IEEE Trans. Pattern Analysis and Machine Intelligence*, 24(10):176–183, 2002.

[27] J. F. Cardoso. High-order contrasts for independent component analysis. *Neural Computation*, 11(1):157–192, Jan. 1999.

[28] H. Chandrasekaran, J. Li, W. H. Delashmit, P. L. Narasimha, C. Yu, and M. T. Manry. Convergent design of piecewise linear neural networks. *Neurocomputing*, 70:1022–1039, 2007.

[29] C. Chatfield and A. J. Collins. *Introduction to Multivariate Analysis*. Chapmanand Hall/CRC, London, 2000.

[30] C. Chatterjee. Adaptive algorithm for first principal eigenvector computation. *Neural Networks*, 18:145–159, 2005.

[31] C. Chatterjee, Z. Kang, and V. P. Roychowdhury. Algorithm for accelerated convergence of adaptive PCA. *IEEE Trans. Neural Networks*, 11(2):338–355, March 2000.

[32] C. Chatterjee, V. P. Roychowdhury, and E. P. Chong. On relative convergence properties of principal component analysis algorithm. *IEEE Trans. Neural Networks*, 9:319–329, 1998.

[33] C. Y. Chee and D. Xu. Chaotic encryption using discrete-time synchronous chaos. *Physics Letters A*, 348:284–292, 2006.

[34] G. Chen, Y. Mao, and C. K. Chui. A sysmetric image encryption based on 3D chaotic maps. *Chaos, Solitons & Fractals*, 21:749–761, 2004.

[35] N. Chen, W. Liu, and J. Feng. Sufficient and necessary condition for the convergence of stochastic approximation algorithms. *Statistics & Probability Letters*, 76(2):203–210, Jan. 2006.

[36] T. Chen and S. Amari. Unified stabilization approach to principal and minor components extraction. *Neural Networks*, 14:1377–1387, 2001.

[37] T. Chen, S. Amari, and Q. Lin. A unified algorithm for principal and minor components extraction. *Neural Networks*, 11:385–390, 1998.

[38] T. Chen, S. Amari, and N. Murata. Sequential extraction of minor components. *Neural Processing Letters*, 13:195–201, 2001.

[39] T. Chen, Y. Hua, and W. Yan. Global convergence of Oja's subspace algorithm for principal component extraction. *IEEE Trans. Neural Networks*, 9(1):58–67, Jan. 1998.

[40] T. Chen and R. W. Liu. An on-line unsupervised learning machine for adaptive feature extraction. *IEEE Trans. Circuits and System-II: Analog and Digital Signal Processing*, 41(1):87–98, 1994.

[41] G. A. Cherry and S. J. Qin. Multiblock principal component analysis based on a combined index for semiconductor fault dectection and diagnosis. *IEEE Trans. Semiconductor Manufacturing*, 19(2):159–172, May 2006.

[42] T. J. Chin and D. Suter. Incremental kernel principal component analysis. *IEEE Trans. Image Processing*, 16(6):1662–1674, Jun. 2007.

[43] A. Cichocki and S. Amari. *Adaptive Blind Signal and Image Processing: Learning Algorithms and Applications.* John Wiley & Sons, Inc., New York, 2002.

[44] A. Cichocki, W. Kasprezak, and W. Skarbek. Adaptive learning algorithm for principal component analysis with partial data. *Proc. Cybernetics Syst.*, 2:1014–1019, 1996.

[45] A. Cichocki and R. Thawonmas. On-line algotithm for blind signal extraction of arbitrarily distributed, but temporally correlated sources using second order statistics. *Neural Processing Letters*, 12(1):91–98, Aug. 2000.

[46] A. Cichocki, R. Thawonmas, and S. Amari. Sequential blind signal extraction in order specified by stochastic properties. *Electronics Letters*, 33(1):64–65, Jan. 1997.

[47] G. Cirrincione. *A Neural Approach to the Structure from Motion Problem.* Ph.D. Dissertation, LIS INPG Grenoble, 1998.

[48] G. Cirrincione, M. Cirrincione, J. Hérault, and S. V. Huffel. The MCA EXIN neuron for the minor cmponent analysis. *IEEE Trans. Neural Networks*, 13(1):160–186, 2002.

[49] P. Comon. Independent component analysis – a new concept? *Signal Processing*, 36:287–314, 1994.

[50] T. Cootes, G. Edwards, and C. Taylor. Active appearance models. In *Proceedings, Volume II. Lecture Notes in Computer Science 1407, ECCV, Freiburg, Germany*, pages 484–498. Springer, 1998.

[51] S. Costa and S. Fiori. Image compression using principal component neural networks. *Image and Vision Computing*, 19:649–668, 2001.

[52] Database for the identification of systems. Available online at: http://www.esat.kuleuven.ac.be/sista/daisy.

[53] J. W. Davis and H. Gao. Gender recognition from walking movements using adaptive three-mode PCA. In *Proceedings of the 2004 IEEE Computing Society Conference on Computer Vision and Pattern Recognition Workshops*, pages 9–16, 2000.

[54] F. De la Torre and M. J. Black. Robust principal component analysis for computer vision. In *Proc. of Int. Conf. on Computer Vision (ICCV'2001), Vancouver, Canada*, 2001.

[55] F. De la Torre and M. J. Black. A framework for robust subspace learning. *International Journal of Computer Vision*, 54(1/2/3):117–142, 2003.

[56] K. I. Diamantaras and S. Y. Kung. *Principal Component Neural Networks (Theory and Application)*. John Wiley & Sons, Inc, New York, 1996.

[57] K. I. Diamantaras and T. Papadimitriou. Subspace-based channel shortening for the blind separation of convolutive mixtures. *IEEE Trans. Signal Processing*, 54(10):3669–3676, Oct. 2006.

[58] M. Doebeli. Intermittent chaos in populaltion dynamics. *Journal of Theoretical Biology*, 166:325–330, 1994.

[59] D. Donoho. *On minimum entropy deconvolution*. Academic Press, New York, 1981.

[60] S. C. Douglas. A self-stabilized minor subspace rule. *IEEE Signal Processing*, 5(12):328–330, 1998.

[61] G. Dror and M. Tsodyks. Chaos in neural networks with dynamical synapses. *Neurocomputing*, 32:365–370, 2000.

[62] W. F. Eddy, W. C. D. Fizgerald, and D. C. Noll. Improved image registration by using Fourier interpolation. *Magent Reson Med*, 36:923–931, 1996.

[63] I. Elhanany, M. Sheinfeld, A. Beck, Y. Kadmon, N. Tal, and D. Tirosh. Robust image registration based on feedforward neural networks. In *Proceedings of 2000 IEEE International Conference on Systems, Man and Cybernetics*, pages 1507–1511, 2000.

[64] E. G. Emberly, R. Mukhopadhyay, C. Tang, and N. S. Wingreen. Flexibility of beta-sheets: principal component analysis of database protein structures. *Proteins*, 55(1):91–98, 2004.

[65] J. Eriksson, J. Karvanen, and V. Koivunen. Source distribution adaptive maximum likelihood estimation of ICA model. In *Proceedings of Second International Workshop on Independent Component Analysis and Blind Signal Separation*, pages 227–232, 2000.

[66] Y. Fand, T. W. S. Chow, and K. T. Ng. Linear neural network based blind equalization. *Signal Processing*, 76:37–42, 1999.

[67] D. Z. Feng, Z. Bao, and L. C. Jiao. Total least mean squares algorithm. *IEEE Trans. Signal Processing*, 46:2122–2130, 1998.

[68] D. Z. Feng, Z. W. Zheng, and Y. Jia. Neural network learning algorithm for tracking minor subspace in high-dimensional data stream. *IEEE Trans. Neural Networks*, 16(3):513–521, 2005.

[69] C. Fookes and M. Bennamoun. The use of mutual information for rigid medical image registration: a review. *Proc. Cybernetics Syst.*, 2:1014–1019, 1996.

[70] E. F. Gad, A. F. Atiya, S. Shaheen, and A. El-Dessouki. A new algorithm for learning in piecewise-linear neural networks. *Neural Networks*, 13:485–505, 2000.

[71] K. Gao, M. O. Ahmad, and M. N Swamy. Learning algorithm for total least squares adaptive signal processing. *Electron. Lett.*, 28(4):430–432, 1992.

[72] K. Gao, M. O. Ahmad, and M. N Swamy. A constrained anti-Hebbian learning algorithm for total least squares estimation with applications to adaptive FIR and IIR filtering. *IEEE Trans. Circuits Syst. Part II*, 41:718–729, 1994.

[73] G. H. Golub and C. F. Van Loan. *Matrix computation.* The Johns Hopkins University Press, Baltimore, 1996.

[74] J. W. Griffiths. Adaptive array processing. In *Proc. Inst. Elect. Eng.*, pages 130:3–10, 1983.

[75] M. T. Hagan and H. B. Demuth. *Neural Network Design.* PWS publishing Company, 1996.

[76] S. Haykin. *Adaptive Filter Theory, third edition.* Prentice Hall, Inc., Englewood Cliffs, 2001.

[77] S. Haykin. *Neural Network, A Comprehensive Foundation, second edition.* Prentice-Hall, Inc., Englewood Cliffs, 2002.

[78] J. H. He. Application of homotopy perturbation method to nonlinear wave equations. *Chaos, Solitons & Fractals*, 26:695–700, 2005.

[79] J. H. He. Limit cycle and bifurcation of nonlinear problems. *Chaos, Solitons & Fractals*, 26:827–833, 2005.

[80] S. Heravi, D. R. Osborn, and C. R. Birchenhall. Linear versus neural network forecasts for European industrial production series. *International Journal of Forecasting*, 20:435–446, 2004.

[81] K. Hornik and C. M. Kuan. Convergence analysis of local feature extraction algorithm. *Neural Networks*, 5:229–240, 1992.

[82] H. Hotelling. Analysis of a complex of statistical variables into principal components. *Journal of Educational Psychology*, 24:417–441, 498–520, 1933.

[83] A. Hyvärinen. Fast and robust fixed-point algorithm for independent component analysis. *IEEE Trans. Neural Networks*, 10(3):626–634, May 1999.

[84] A. Hyvärinen, J. Karhunen, and E. Oja. *Independent Component Analysis*. John Wiley & Sons, Inc., New York, 2001.

[85] A. Hyvärinen and E. Oja. A fast fixed-point algorithm for independent component analysis. *Neural Computation*, 9(7):1483–1492, 1997.

[86] A. Hyvärinen and E. Oja. Simple neuron models for independent component analysis. *International Journal of Neural Systems*, 7(6):671–687, 1997.

[87] A. Hyvärinen and E. Oja. Independent component analysis by general non-linear Hebbian-like learning rules. *Signal Processing*, 64(3):301–313, 1998.

[88] A. Hyvärinen and E. Oja. Independent component analysis: algorithms and applications. *Neural Networks*, 13(4–5):411–430, 2000.

[89] M. G. Jafari and J. A. Chambers. Fetal electrocardiogram extraction by sequential source separation in the wavelet domain. *IEEE Trans. Biomedical Engineering*, 52(3):390–400, Mar. 2005.

[90] S. M. Jakubek and T. I. Stasser. Artificial nerual networks for fault detection in large-scale data acquisition system. *Engineering Applications of Artificial Intelligence*, 17:233–248, 2004.

[91] M. V. Jankovic and H. Ogawa. Modulated Hebb-Oja learning rule–a method for principal subspace analysis. *IEEE Trans. Neural Networks*, 17(2):345–356, Mar. 2006.

[92] Z. Jing and J. Yang. Bifurcation and chaos in discrete-time predator-prey system. *Chaos, Solitons & Fractals*, 27:259–277, 2006.

[93] I. J. Jolliffe. *Principal Component Analysis, second edition*. Springer-verlag, New York, Berlin, Heidelberg, 2002.

[94] C. Jutten and J. Herault. Blind separation of sources, Part I: an adaptive algorithm based on neuromimetic architecture. *Signal Processing*, 24:1–10, 1991.

[95] P. P. Kanjilal and S. Palit. Fetal ECG extraction from single-channel maternal ECG using singular value decomposition. *IEEE Trans. Biomedical Engineering*, 44(1):51–59, Jan. 1997.

[96] J. Karhunen, A. Hyvärinen, R. Vigario, J. Hurri, and E. Oja. Applications of neural blind separation to signal and image processing. In *Proc. IEEE Int. Conf. on Acoustics, Speech and Signal Processing (ICASSP 97)*, pages 131–134, 1997.

[97] J. Karhunen, E. Oja, L. Wang, R. Vigario, and J. Joutsensalo. A class of neural networks for independent component analysis. *IEEE Trans. Neural Networks*, 8(3):486–504, 1997.

[98] J. Karvanen, J. Eriksson, and V. Koivunen. Pearson system based method for blind separation. In *Proceedings of Second International Workshop on Independent Component Analysis and Blind Signal Separation*, pages 585–590, 2000.

[99] K. I. Kim, M. O. Franz, and B. Schölkopf. Iterative kernel principal component analysis for image modeling. *IEEE Trans. Pattern Analysis and Machine Intelligence*, 27(9):1351–1366, Sep. 2005.

[100] R. Klemm. Adaptive airborne MTI: An auxiliary channel approach. In *Proc. Inst. Elect. Eng.*, pages 134:269–276, 1987.

[101] T. Kohonen. The self-organizing map. *Neurocomputing*, 21:1–6, 1998.

[102] C. J. Ku and T. L. Fine. Testing for stochastic independence: application to blind source separation. *IEEE Trans. Signal Processing*, 53(5):1815–1826, May 2005.

[103] S. Y. Kung, K. I. Diamantaras, and J. S. Taur. Adaptive principal component extraction (APEX) and applications. *IEEE Trans. Signal Processing*, 42(5):1202–1217, May 1994.

[104] N. Kwak. Principal component analysis based on L1-norm maximization. *IEEE Trans. Pattern Analysis and Machine Intelligence*, 30(9):1672–1680, Sep. 2008.

[105] C. Kyrtsou and M. Terraza. Is it possible to study chaotic and ARCH behaviour jointly? application of a noisy Mackey-Glass equation with heteroskedastic errors to the Paris Stock Exchange returns series. *Computational Economics*, 21:257–276, 2003.

[106] L. D. Lathauwer, J. Castaing, and J. F. Cardoso. Fourth-order cumulant-based blind identification of underdetermined mixtures. *IEEE Trans. Signal Processing*, 55(6):2965–2973, Jun. 2007.

[107] L. D. Lathauwer, B. D. Moor, and J. Vandewalle. Fetal electrocardiogram extraction by blind source subspace separation. *IEEE Trans. Biomedical Engineering*, 47(5):567–572, May 2000.

[108] L. D. Lathauwer, B. D. Moor, and J. Vandewalle. Independent component analysis and (simultaneous) third-order tensor diagonalization. *IEEE Trans. Signal Processing*, 49(10):2262–2271, Oct. 2001.

[109] J. M. Le, C. Yoo, and I. B. Lee. Fault detection of batch processes using multiway kernel principal component analysis. *Computers and Chemical Engineering*, 28:1837–1847, 2004.

[110] C. Li and G. Chen. Chaos in the fractional order Chen system and its control. *Chaos, Solitons & Fractals*, 22:549–554, 2004.

[111] Y. Li and Z. Yi. An algorithm for extracting fetal electrocardiogram. *Neurocomputing*, 71:1538–1542, 2008.

[112] H. Lian. Bayesian nonlinear principal component analysis using random fields. *IEEE Trans. Pattern Analysis and Machine Intelligence*, 31(4):749–754, Apr. 2009.

[113] C. T. Lin, S. A. Chen, C. H. Huang, and Chung J. F. Cellular neural networks and PCA neural networks based rotation/scale invariant texture classification. In *Proceedings of 2004 IEEE International Joint Conference on Neurl Networks*, pages 153–158, 2004.

[114] C. Liu. Gabor-based kernel PCA with fractional power polynomial models for face recogniton. *IEEE Trans. Pattern Analysis and Machine Intelligence*, 26(5):572–581, May 2004.

[115] W. Liu and D. P. Mandic. A normalised Kurtosis based blind source extraction from noisy mixture. *Signal Processing*, 86(7):1580–1585, Jul. 2006.

[116] Z. Y. Liu and L. Xu. Topological local principal component analysis. *Neurocomputing*, 55:739–745, 2003.

[117] L. Ljung. Analysis of recursive stochastic algorithms. *IEEE Trans. Automat. Contr.*, 22(4):551–575, Aug. 1977.

[118] W. Z. Lu, W. J. Wang, X. K. Wang, S. H. Yan, and J. C. Lam. Potential assessment of a neural network model with PCA/RBF approach for forecasting pollutant trends in Mong Kok urban air, Hong Kong. *Environmental Research*, 96:79–87, 2004.

[119] F. Luo, R. Unbehauen, and A. Cichocki. A minor component analysis algorithm. *Neural Networks*, 10:291–297, 1997.

[120] J. C. Lv, K. K. Tan, Z. Yi, and S. Huang. Convergence analysis of a class of Hyvärinen-Oja's ICA learning algorithms with constant learning rates. *IEEE Trans. Signal Processing*, 10(5):1811–1824, May 2008.

[121] J. C. Lv, K. K. Tan, Z. Yi, and S. Huang. Stability and chaos of a class of learning algorithm for ICA neural networks. *Neural Processing Letters*, 28(1):35–47, Aug. 2008.

[122] J. C. Lv and Z. Yi. Some chaotic behaviors in a MCA learning algorithm. *Chaos Solitions & Fractals*, 33:1040–1047, 2007.

[123] J. C. Lv and Z. Yi. Stability and chaos of LMSER PCA learning algorithm. *Chaos Solitions & Fractals*, 32:1440–1447, 2007.

[124] J. C. Lv, Z. Yi, and K. K. Tan. Convergence analysis of Xu's LMER learning algorithm via deterministic discrete time system method. *Neurocomputing*, 70:362–372, 2006.

[125] J. C. Lv, Z. Yi, and K. K. Tan. Global convergence of Oja's PCA learning algorithm with a non-zero-approaching adaptive learning rate. *Theoretical Computer Science*, 367:286–307, 2006.

[126] J. C. Lv, Z. Yi, and K. K. Tan. Determining of the number of principal directions in a biologically plausible PCA model. *IEEE Trans. Neural Networks*, 18(2):910–916, 2007.

[127] J. C. Lv, Z. Yi, and K. K. Tan. Global convergence of GHA learning algorithm with nonzero-approaching learning rates. *IEEE Trans. Neural Networks*, 18(6):1557–1571, 2007.

[128] G. Mathew and V. Reddy. Development and analysis of a neural network approach to Pisarenko's harmonic retrieval method. *IEEE Trans. Signal Processing*, 42:663–667, 1994.

[129] W. B. Mikhael and T. Yang. A gradient-based optimum block adaption ICA technique for interference suppression in highly dynamic communication channels. *EURASIP Journal on Applied Signal Processing*, pages 1–10, 2006.

[130] R. Möller. A self-stabilizing learning rule for minor component analysis. *International Journal of Neural Systems*, 14(1):1–8, 2004.

[131] R. Möller and A. Könies. Coupled principal component analysis. *IEEE Trans. Neural Networks*, 15(1):214–222, 2004.

[132] M. G. Mostafa, A. A. Farag, and E. Essock. Multimodality image registration and fusion using neural network information. In *Proceedings of the Third International Conference on Information Fusion, Paris, France*, pages 10–13, 2000.

[133] K. Nishino, S. K. Nayar, and T. Jebara. Clustered blockwise PCA for representing visual data. *IEEE Trans. Pattern Analysis and Machine Intelligence*, 27:1675–1679, 2005.

[134] E. Oja. A simplified neuron mode as a principal component analyzer. *J. Math. Biol.*, 15:167–273, 1982.

[135] E. Oja. Principal components, minor components, and linear neural networks. *Neural Networks*, 5:927–935, 1992.

[136] E. Oja and J. Karhunen. On stochastic approximation of the eigenvector and eigenvalues of the expectation of a random matrix. *J. Math. Anal. App.*, 106(1):69–84, Feb. 1985.

[137] E. Oja and L. Wang. Robust fitting by nonlinear neural units. *Neural Networks*, 9(3):435–444, 1996.

[138] E. Oja and Z. Yuan. The fastICA algorithm revisited: convergence analysis. *IEEE Trans. Neural Networks*, 17(6):1370–1381, 2006.

[139] S. Ouyang, Z. Bao, and G. S. Liao. Robust recursive least squares learning algorithm for principal component analysis. *IEEE Trans. Neural Networks*, 11(1):215–221, Jan. 2000.

[140] N. K. Pareek, V. Patidar, and K. K. Sud. Image encryption using chaotic logistic map. *Image and Vision Computing*, 24:926–934, 2006.

[141] K. Pearson. On lines and planes of closest fit to systems of points in space. *Philosophical Magazine*, 2(6):559–572, 1901.

[142] D. Peng and Z. Yi. Convergence analysis of the OJAn MCA learning algorithm by the deterministic discrete time method. *Theoretical Computer Science*, 378(1):87–100, 2007.

[143] D. Peng and Z. Yi. Dynamics of generalized PCA and MCA learning algorithms. *IEEE Trans. Neural Networks*, 18(6):1777–1784, 2007.

[144] D. Peng and Z. Yi. A modified Oja-Xu MCA learning algorithm and its convergence analysis. *IEEE Trans. Circuits and Systems II*, 54(4):348–352, 2007.

[145] D. Peng, Z. Yi, and J. C. Lv. A stable MCA learning algorithm. *Computer & Mahthematics with Applications*, 56(4):847–860, 2008.

[146] D. Piraino, P. Kotsas, and M. Recht. Three dimensional image registration using artificial neural networks. In *Proceedings of IEEE International Conference on Neural Networks*, pages 4017–1021, 2003.

[147] M. D. Plumbley and E. Oja. A "Nonnegative PCA" algorithm for independent component analysis. *IEEE Trans. Neural Networks*, 15(1):66–76, 2004.

[148] A. Pujol, J. Vitrià, F. Lumbreras, and J. J. Villanueva. Topological principal component analysis for face encoding and recognition. *Pattern Recognition Letters*, 22(6-7):769–776, May 2001.

[149] P. A. Regalia and E. Kofidis. Monotonic convergence of fixed-point algorithms for ICA. *IEEE Trans. Neural Networks*, 14(4):943–949, July 2003.

[150] S. T. Roweis and L. K. Saul. Nonlinear dimensionality reduction by locally linear embedding. *Science*, 290:2323–2326, 2000.

[151] E. L. Rubio, O. L. Lobato, J. M. Pérez, and J. A. G. Ruiz. Principal component analysis competitive learning. *Neural Computation*, 16:2459–2481, 2004.

[152] B. G. Ruessink, I. M. J. Enckevort, and Y. Kuriyama. Non-linear principal component analysis of nearshore bathmetry. *Marine Geology*, 23:185–197, 2004.

[153] I. Sadinezhad and M. Joorabian. A novel frequency tracking method based on complex adaptive linear neural network state vector in power systems. *Electric Power System Research*, 79:1216–1225, 2009.

[154] H. Sakai and K. Shimizu. A new adaptive algorithm for minor component analysis. *Signal Processing*, 71:301–308, 1998.

[155] T. H. Sander, M. Brughoff, G. Gurio, and L. Trahms. Single evoked somatosensory MEG responses extracted by time delayed decorrelation. *IEEE Trans. Signal Processing*, 53(9):3384–3392, Sep. 2005.

[156] T. D. Sanger. Optimal unsupervised learning in a single-layer linear feedforward neural network. *Neural Networks*, 2:459–473, 1989.

[157] R. Schmidt. Multiple emitter location and signal parameter estimation. *IEEE Trans. Autennas Propagation*, 34:276–280, 1986.

[158] A. Selamat and S. Omatu. Web page feature selection and classification using neural networks. *Information Sciences*, 158:69–88, 2004.

[159] O. Shalvi and E. Weinstein. New criteria for blind deconvolution of non-minimum phase systems (channels). *IEEE Trans. Information Theory*, 36:312–321, 1990.

[160] L. Shang, J. C. Lv, and Z. Yi. Rigid medical image registration using PCA neural networks. *Neurocomputing*, 69:1717–1722, 2006.

[161] J. E. Skinner, M. Molnar, T. Vybiral, and M. Mitra. Application of chaos theory to biology and medicine. *Integrative Psychological and Behavioral Science*, 27(1):39–53, Jan. 1992.

[162] V. Solo and X. Kong. Performance analyisis of adaptive eigenanalysis algorithms. *IEEE Trans. Signal Processing*, 46:636–646, Mar. 1998.

[163] P. Stavroulakis. *Chaos Applications in Telecommunications*. CRC Press, Inc., Boca Raton, 2005.

[164] J. V. Stone. *Independent Component Analysis: A Tutorial Introduciton*. The MIT Press, 2004.

[165] J. A. K. Suykens, T. V. Gestel, J. Vandewalle, and B. Demoor. A support vector machine formulation to PCA analysis and its kernel version. *IEEE Trans. Neural Networks*, 14(2):447–450, 2003.

[166] K. Tan and S. Chen. Adaptively weighted sub-pattern PCA face recognition. *Neurocomputing*, 64:505–511, 2005.

[167] Q. Tao, X. Liu, and X. Cui. A linear optimization neural network for associative memory. *Applied Mathematics and Computation*, 171:1119–1128, 2005.

[168] J. Thomas, Y. Deville, and S. Hosseini. Differential fast fixed-point algorithms for underdetermined instantaneous and convolutive partial blind source separation. *IEEE Trans. Signal Processing*, 55(7):3717–3729, Jul. 2007.

[169] J. Tichavský, Z. Koldovský, and E. Oja. Performance analysis of the fastICA algorithm and Cramér-Rao bounds for linear independent component analysis. *IEEE Trans. Signal Processing*, 54(4):1189–1203, Apr. 2006.

[170] M. Tipping and C. Bishop. Mixtures of probabilistic principal component analyzers. *Neural Computation*, 11(2):443–482, 1999.

[171] M. Tipping and C. Bishop. Probabilistic principal component analysis. *J. Royal Statistical Soc.*, 61(3):611–622, 1999.

[172] J. M. Vegas and P. J. Zufiria. Generalized neural networks for spectral analysis: dynamics and Liapunov functions. *Neural Networks*, 17:233–245, 2004.

[173] P. Vidal, Y. Ma, and S. Sastry. Generalized principal component analysis (GPCA). *IEEE Trans. Pattern Analysis and Machine Intelligence*, 27(12):1945–1959, Dec. 2005.

[174] V. Vigneron, A. Paraschiv-Lonescu, A. Azancot, C. Jutten, and O. Sibony. Fetal electrocardiogram extraction based on non-stationary ICA and wavelet denoising. In *Seventh International Symposium on Signal Processing and Its Applications, Paris, France*, pages 69–72, 2003.

[175] T. Voegtlin. Recursive principal component analysis. *Neural Networks*, 18:1051–1063, 2005.

[176] V. Šmídl and A. Quinn. On Bayesian principal component analysis. *Computational Statistics & Data Analysis*, 51:4101–4123, 2007.

[177] R. Wan and M. Li. An overview of medical image registration. In *Proceedings of the Fifth International Conference on Computational Interlligence and Multivedia Application (ICCMA 2003)*, pages 385–390, 2003.

[178] X. Wang and L. Tian. Bifurcation analysis and linear control of the Newton-Leipnik system. *Chaos, Solitons & Fractals*, 27:31–38, 2006.

[179] A. Weingessel and K. Hornik. Local PCA algorithms. *IEEE Trans. Nerual Networks*, 11(6):1242–1250, Jan. 2000.

[180] C. Wen and X. Ma. A max-piecewise-linear neural network for function approximation. *Neurocomputing*, 71:843–852, 2008.

[181] B. Widrow and M. E. Hoff. Adaptive switching circuits. *IRE WESCON Convention Record, New York: IRE Part 4*, pages 96–104, 1960.

[182] F. Wu and Qian Z. C. Principal axes algorithm based on image contour applied to medical image registration. *J. Fourth Mil. Med. U.*, 22:567–569, 2001.

[183] F. Xiao, G. Liu, and B. Zheng. A chaotic encryption system using PCA neural networks. In *Proceedings of IEEE Conference on Cybernetics and Intelligent Systems*, pages 465–469, 2008.

[184] L. Xu. Least mean square error reconstruction principle for self-organizing neuro-nets. *Neural Networks*, 6:627–648, 1993.

[185] L. Xu, E. Oja, and C. Y. Suen. Modified Hebbian learning for curve and surface fitting. *Nerual Networks*, 5:441–457, 1992.

[186] M. Ye. Global convergence analysis of a self-stabilized MCA learning algorithm. *Neurocomputing*, 67:321–327, July 2005.

[187] M. Ye. Global convergence analysis of a discrete time nonnegative ICA algorithm. *IEEE Trans. Neural Networks*, 17(1):253–256, Jan. 2006.

[188] M. Ye, X. Q. Fan, and X. Li. A class of self-stabilizing MCA learning algorithms. *IEEE Trans. Neural Networks*, 17(5):1634–1636, 2006.

[189] M. Ye, Z. Yi, and J. C. Lv. A globally convergent learning for PCA neural networks. *Neural Computing & Application*, 14:18–24, 2005.

[190] Z. Yi, Y Fu, and H. J. Tang. Neural networks based approach computing eigenvectors and eigenvalues of symmetric matrix. *Computers and Mathematics with Applications*, 47:1155–1164, 2004.

[191] Z. Yi and K. K. Tan. Dynamical stability conditions for Lotka-Volterra recurrent neural networks with delays. *Neural Computation*, 66(1):011910, Jul. 2002.

[192] Z. Yi and K. K. Tan. *Convergence Analysis of Recurrent Neural Networks*. Kluwer Academic Publishers, Boston, 2004.

[193] Z. Yi and K. K. Tan. Multistability analysis of discrete recurrent neural networks with unsaturating piecewise linear transfer functions. *IEEE Trans. Neural Networks*, 15(2):329–336, Mar. 2004.

[194] Z. Yi, K. K. Tan, and T. H. Lee. Multistability analysis for recurrent neural networks with unsaturating piecewise linear transfer functions. *Neural Computation*, 15:639–662, Mar. 2003.

[195] Z. Yi, M. Ye, J. C. Lv, and K. K. Tan. Convergence analysis of a deterministic discrete time system of Oja's PCA learning algorithm. *IEEE Trans. Neural Networks*, 16(6):1318–13128, Nov. 2005.

[196] C. Yin, Y. Shen, S. Liu, Q. Yin, W. Guo, and Z. Pan. Simultaneous quantitative UV spectrophotometric determination of multicomponents of amino acids using linear neural network. *Computer and Chemistry*, 25:239–243, 2001.

[197] V. Zarzoso and P. Comon. How fast is fastICA? In *Proc. EUSIPCO-2006, XIV European Signal Processing Conference, Florence, Italy*, pages 2–4, 2006.

[198] V. Zarzoso and P. Comon. Comparative speed analysis of fastICA. In *Proc. ICA-2007, 7th Internaltional Conference on Independent Component Analysis and Signal Separation, London, UK, Lecture Notes in Computer Science*, pages 9–21. Springer, Berlin, 2007.

[199] V. Zarzoso, J. J. Murillo-Fuentes, R. Boloix-Tortosa, and A. K. Nandi. Optimal pairwise fourth-order independent component analysis. *IEEE Trans. Signal Processing*, 54(8):3049–3063, Aug. 2006.

[200] V. Zarzoso and A. K. Nandi. Noninvasive fetal electrocardiogram extraction blind separation versus adaptive noise cancellation. *IEEE Trans. Biomedical Engineering*, 48(1):12–18, Jan. 2001.

[201] Q. Zhang. Global analysis of Oja's flow for neural networks. *Neurocomputing*, 55:761–769, Oct. 2003.

[202] Q. Zhang. On the discrete-time dynamicas of a PCA learning algorithm. *Neurocomputing*, 55:761–769, 2003.

[203] Q. Zhang and Y. W. Leung. Energy function for one-unit Oja algorithm. *IEEE Trans. Neural Networks*, 6(5):1291–1293, Sept. 1995.

[204] Q. Zhang and Y. W. Leung. A class of learning algorithms for principal component analysis and minor component analysis. *IEEE Trans. Neural Networks*, 11(2):529–533, Mar. 2000.

[205] Z. L. Zhang and Z. Yi. Extraction of a source signal whose kurtosis value lies in a specific range. *Neurocomputing*, 69(7–9):900–904, Mar. 2006.

[206] Z. L. Zhang and Z. Yi. Robust extraction of specific signals with temporal structure. *Neurocomputing*, 69(7-9):888–893, Mar. 2006.

[207] J. Zhao and Q. Jiang. Probabilistic PCA for t distribution. *Neurocomputing*, 69:2217–2226, 2006.

[208] H. Zhou, T. Hastie, and R. Tibshirani. Sparse principal component analysis. *Journal of Computational and Graphical Statistics*, 15(2):265–286, 2006.

[209] M. Zibulevsky. Extraction of a source from multichannel data using sparse decomposition. *Neurocomputing*, 49(1–4):163–173, Dec. 2002.

[210] P. J. Zufiria. On the discrete-time dynamics of the basic Hebbian neural-network nods. *IEEE Trans. Neural Networks*, 13(6):1342–1352, Nov. 2002.

Index